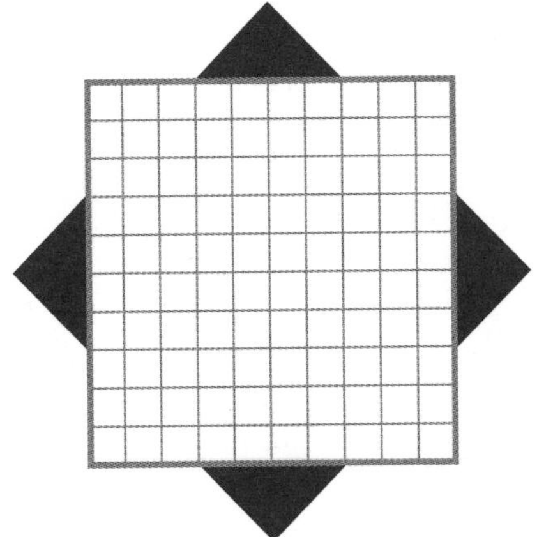

1–100 Activity Book

Activities and Worksheets for the Hundred Board

Dawn Hickman Bacarella

Illustrations by: Tirza Ernst
Design by: Ellen Hart/Concepts Plus, Inc.

ISBN: 1-56911-922-8

Printed in the United States of America.

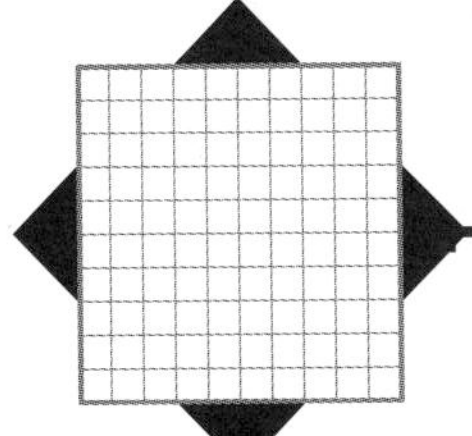

Table of Contents

The Hundred Board

Readiness Activities

Order and Comparison

Place Value Activities

Number Patterns

Addition and Subtraction

Multiplication

Introduction

1-100 Activity Book has been designed in conjunction with the Laminated Hundred Board (LER 0375), but may be used independently. Children will write on their hundred boards for selected activities. Teachers may choose to laminate a class-room set that can be reused for various activities or placed in learning centers.

This book contains more than 66 teacher-directed activities and is divided into six main sections:

- Readiness Activities
- Order and Comparison
- Place Value
- Number Patterns
- Addition and Subtraction
- Multiplication

Each section contains detailed Teacher's Notes and clear, reproducible worksheets. Use the activities and games to introduce and support instructional material detailed in the Table of Contents.

When a worksheet supports an activity, it is outlined in the Teacher's Notes.

The materials needed for each activity are listed in the Teacher's Notes, including:

- Hundred Board worksheets
  Blank Hundred Board (page 1)
  Numbered Hundred Board (page 2)
- Transparent counters in red, blue, yellow, and green (LER 0374)
- Opaque counters in red, blue, yellow, and green (LER 193)
- Number tiles, 1-100 (0381)

This equipment can be purchased or made from cardboard. Laminate homemade materials for a longer shelf-life.

After introducing concepts, encourage children to work in a variety of arrange-ments, including cooperative groups, partners, and individually.

The first two pages of this book are reproducible hundred boards (blank and numbered). Turn them into overhead transparencies when introducing new games and activities to the entire class.

Blank Hundred Board

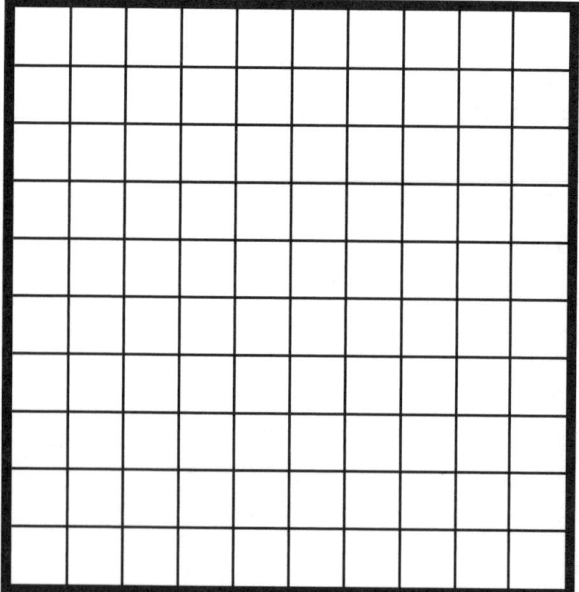

Numbered Hundred Board

1	2	3	4	5	6	7	8	9	10
11	12	13	14	15	16	17	18	19	20
21	22	23	24	25	26	27	28	29	30
31	32	33	34	35	36	37	38	39	40
41	42	43	44	45	46	47	48	49	50
51	52	53	54	55	56	57	58	59	60
61	62	63	64	65	66	67	68	69	70
71	72	73	74	75	76	77	78	79	80
81	82	83	84	85	86	87	88	89	90
91	92	93	94	95	96	97	98	99	100

Name ______________________________

Blank Hundred Board

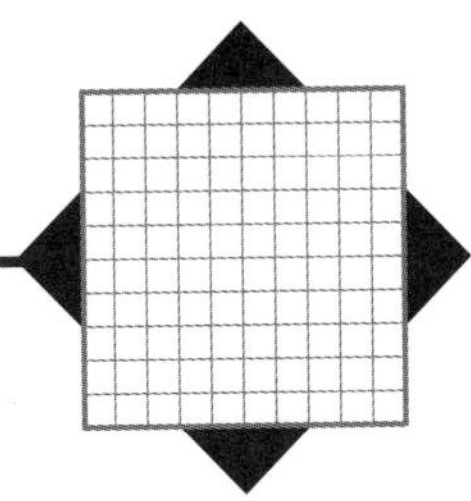

Name ______________________________

Numbered Hundred Board

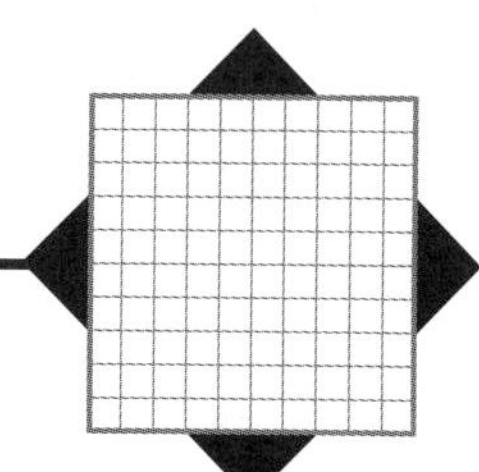

1	2	3	4	5	6	7	8	9	10
11	12	13	14	15	16	17	18	19	20
21	22	23	24	25	26	27	28	29	30
31	32	33	34	35	36	37	38	39	40
41	42	43	44	45	46	47	48	49	50
51	52	53	54	55	56	57	58	59	60
61	62	63	64	65	66	67	68	69	70
71	72	73	74	75	76	77	78	79	80
81	82	83	84	85	86	87	88	89	90
91	92	93	94	95	96	97	98	99	100

Teacher's Notes

Activity 1

The Hundred Board

Materials
- Blank Hundred Board
- 20-30 counters in assorted colors

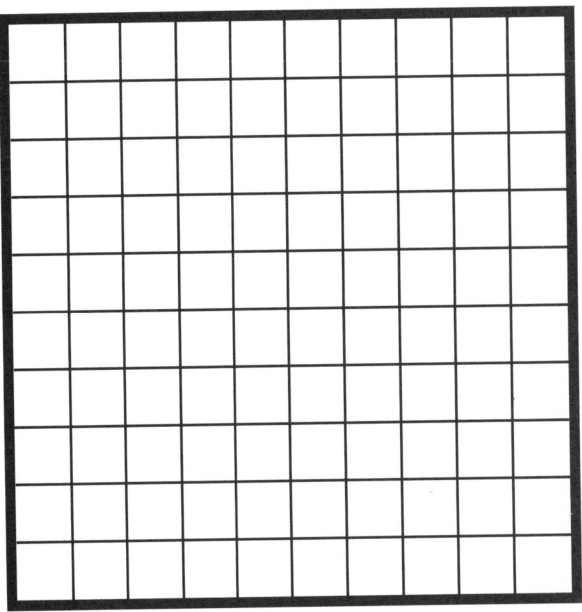

Using the Blank Hundred Board, challenge students to arrange counters to form patterns. Encourage them to create letters, shapes, faces, and other patterns.

Ask students to sort their piles of counters by color, placing each color in a different row on the Blank Hundred Board. Ask them to count the colors in each row. Which color appears most and least often? Do any of the rows have the same number of counters?

Activity 2

Patterns

Materials
- Blank Hundred Board
- 50-60 counters in assorted colors
- Worksheet 2: Pattern Drawings

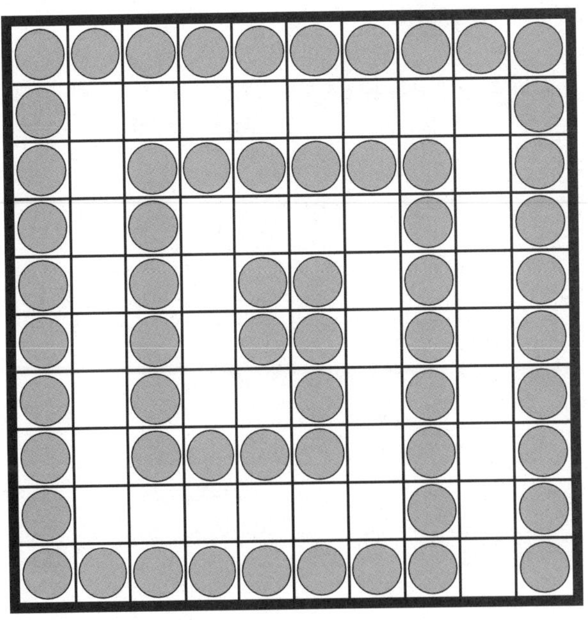

Working in pairs, ask students to alternate between two colors of counters to create a pattern on the hundred board. Challenge children to add three or four more colors to create more patterns. Partners can try to predict and complete each other's patterns.

On the right is an example of a pattern drawing on a hundred board. Another example is shown on Worksheet 2 (page 7).

Activity 3

Making A Graph

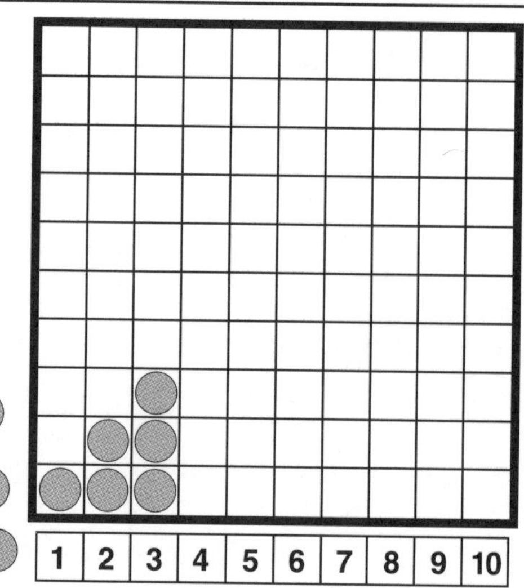

Materials
- Blank Hundred Board
- Number tiles 1-10
- 55 counters

Using the Blank Hundred Board, ask children to place number tiles 1-10, in order, under the bottom row. Then, ask them to place the correct number of counters above each number to create a graph.

Activity 4

Graphing Even and Odd Numbers

Materials
- Blank Hundred Board
- Number tiles 1-10
- 55 counters

Ask students to make graphs of even and odd numbers (use the same method as Activity 3). For an even graph, use number tiles 2, 4, 6, 8, and 10. For an odd graph, use number tiles 1, 3, 5, 7, and 9.

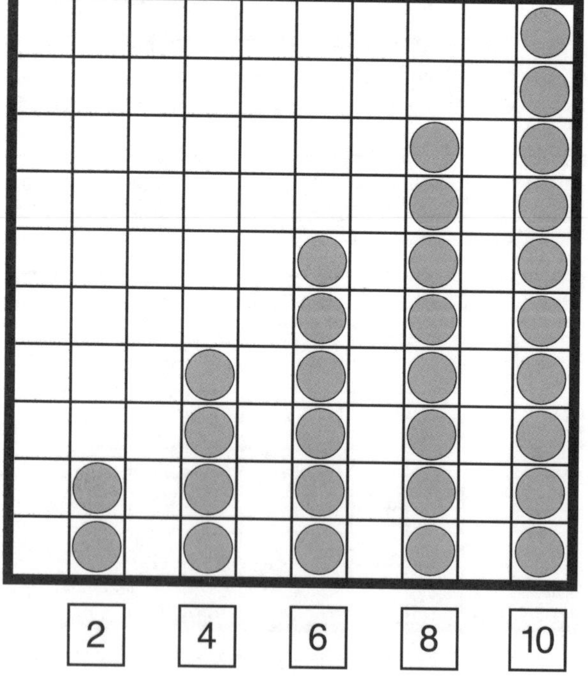

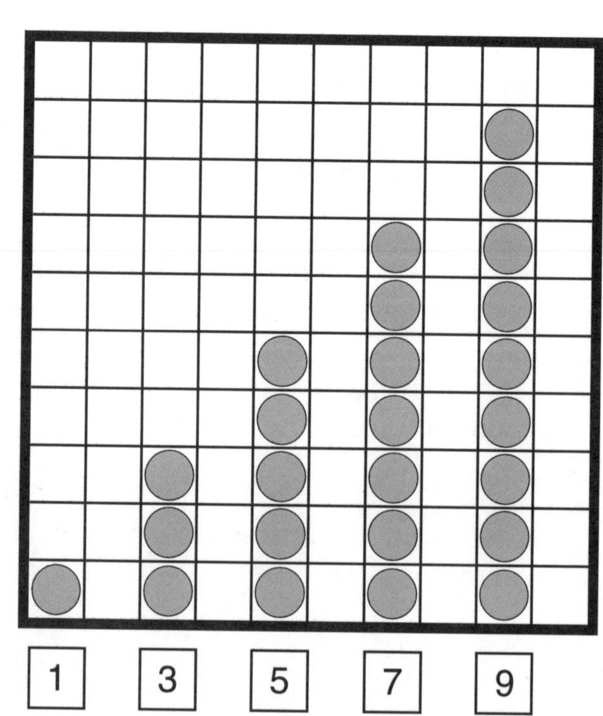

Teacher's Notes

Activity 5

Graphing Data

Materials
- Blank Hundred Board
- Green and blue counters

Ask children to create graphs comparing various traits (hair color, eye color, shoe size, etc.). Use green counters to represent girls and blue counters to represent boys.

Hair Color

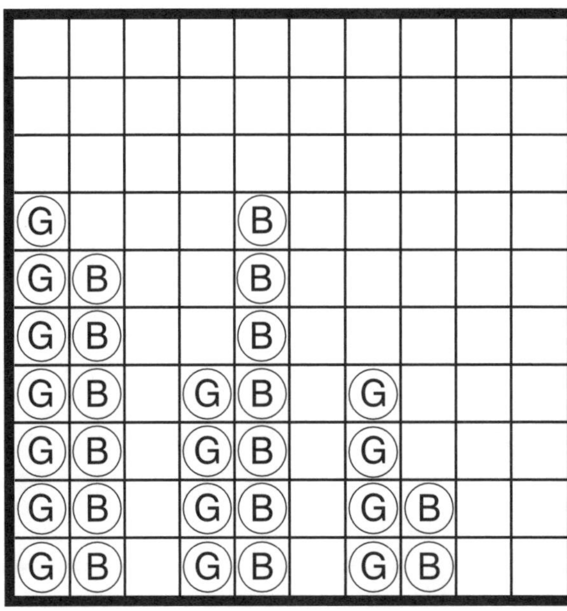

Brown Black Blond

Activity 6

Highest and Lowest

Materials
- Numbered Hundred Board
- Worksheet 6: Highest and Lowest
- Blank Hundred Board
- Number tiles 1-100

Introduce children to the hundred board by discussing rows, columns, digits, counting, and skip-counting. Count through the numbers together. Then give each student a copy of Worksheet 6 and a Numbered Hundred Board to complete. Lastly, ask children to place number tiles, in order, on the Blank Hundred Board.

Numbered Hundred Board

1	2	3	4	5	6	7	8	9	10
11	12	13	14	15	16	17	18	19	20
21	22	23	24	25	26	27	28	29	30
31	32	33	34	35	36	37	38	39	40
41	42	43	44	45	46	47	48	49	50
51	52	53	54	55	56	57	58	59	60
61	62	63	64	65	66	67	68	69	70
71	72	73	74	75	76	77	78	79	80
81	82	83	84	85	86	87	88	89	90
91	92	93	94	95	96	97	98	99	100

Teacher's Notes

Activity 7

What Have You Drawn?

Materials • Numbered Hundred Board

Instruct children to follow the directions to create an illustration. (They will create a house.)

1	2	3	4	5	6	7	8	9	10
11	12	13	14	15	16	17	18	19	20
21	22	23	24	25	26		28	29	30
31	32	33	34	35	36	37	38	39	40
41	42	43	44	45	46	47	48	49	50
51	52	53	54	55	56	57	58	59	60
61	62		64	65		67		69	70
71	72	73	74	75		77	78	79	80
81	82	83	84	85	86	87	88	89	90
91	92	93	94	95	96	97	98	99	100

- Draw a dot in the middle of square 23 and square 28. Use a pencil and ruler to join these dots.
- Draw a dot in the middle of square 42 and square 49. Join these dots with a straight line.
- Join dot 42 to dot 23. Then match dot 28 to dot 49.
- Draw a dot in the middle of square 82 and square 89. Join dot 82 to dot 89; dot 42 to dot 82; and dot 49 to dot 89.
- Color squares 27, 63, 66, 68, 76, and the top half of square 86.

Activity 8

Number Patterns

Materials • Numbered Hundred Board
• 50 counters
• Worksheet 8: Number Patterns

Use a variety of colored counters to show the following patterns on the Numbered Hundred Board. Be sure to use different colors for each pattern.

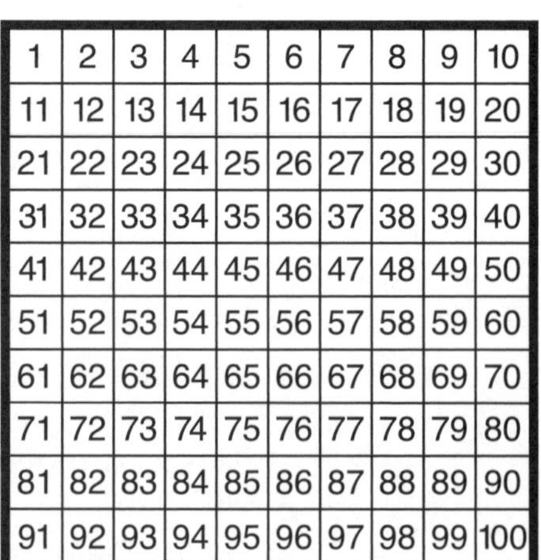

- Count in twos starting with 4.
- Count in fives starting with 25.
- Start on square 7. Add 9 and color the answer. Add 9 six times, coloring each answer.

Point out the visual patterns that emerge. Complete Worksheet 8 for more practice.

Name ____________________

Pattern Drawings

Getting Ready

Worksheet 2

With counters, copy the pattern shown in the first box. Then make your own patterns in the three blank boxes. Don't forget to color the patterns.

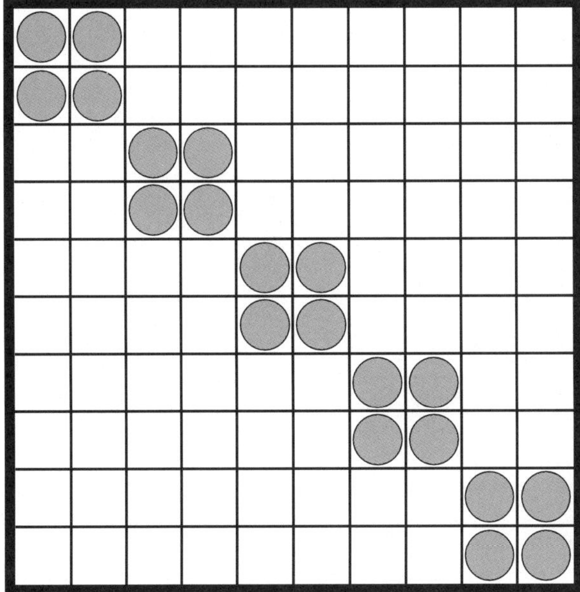

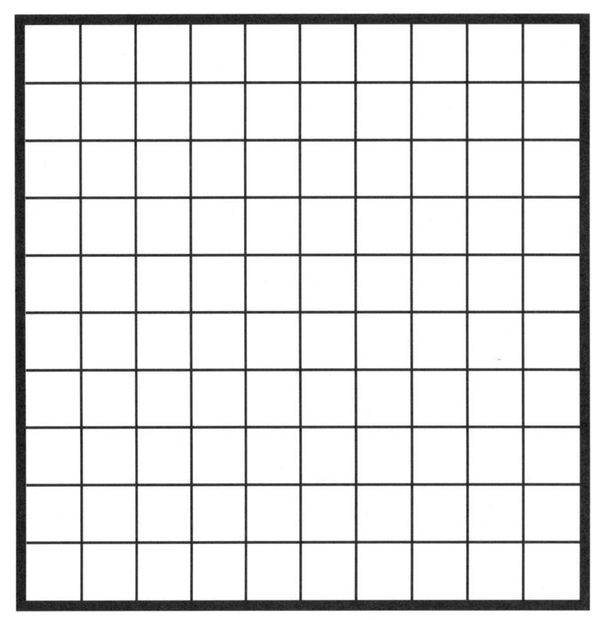

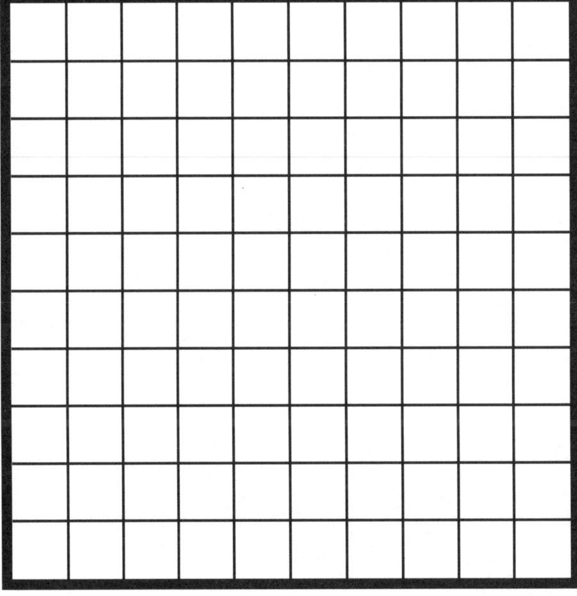

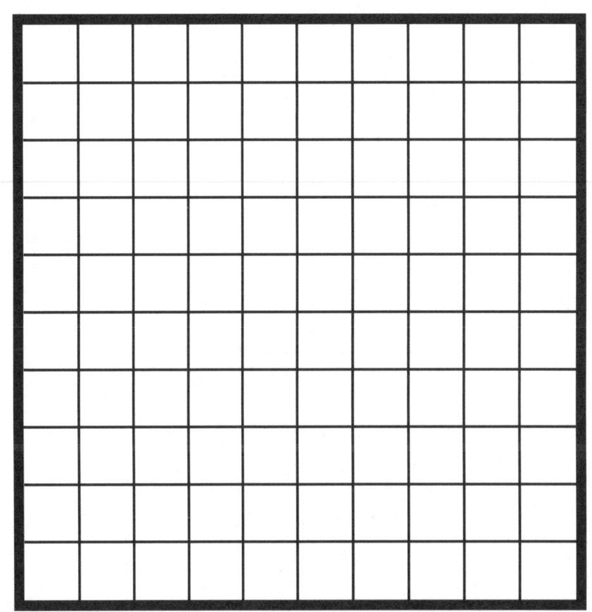

Name ______________________________

Highest and Lowest

Getting Ready

Worksheet 6

1. How many numbers in each row? _____

2. How many rows of numbers? _____

3. Write the lowest number. _____

4. Write the highest number. _____

5. Write the numbers in the 2nd row.

_____ _____ _____ _____ _____ _____ _____ _____ _____ _____

6. Which numbers end with zero?

_____ _____ _____ _____ _____ _____ _____ _____ _____ _____

7. Which numbers have just 1 digit?

_____ _____ _____ _____ _____ _____ _____ _____ _____

8. Which number is at the end of the 2nd row? _____

9. Which number is at the start of the 7th row? _____

10. Which number is the lowest 2-digit number? _____

11. Which number is the highest 2-digit number? _____

Name ______________________________

Number Patterns

Getting Ready

Worksheet 8

Use counters to show these patterns on your hundred board. Then, write the missing numbers.

1. 2, 12, 22, _____ , _____ , _____ , _____ , _____ , _____ , _____

2. 8, 18, 28, _____ , _____ , _____ , _____ , _____ , _____ , _____

3. 11, 22, 33, _____ , _____ , _____ , _____ , _____ , _____

4. 9, 18, 27, _____ , _____ , _____ , _____ , _____ , _____

5. 20, 25, 30, _____ , _____ , _____ , _____ , _____ , _____ , _____

6. 4, 15, 26, _____ , _____ , _____ , _____

7. 7, 16, 25, _____ , _____ , _____ , _____

8. 5, 10, 15, _____ , _____ , _____ , _____ , _____ , _____ , _____

Make up some patterns of your own.

9. _____ , _____ , _____ , _____ , _____ , _____ , _____

10. _____ , _____ , _____ , _____ , _____ , _____ , _____

11. _____ , _____ , _____ , _____ , _____ , _____ , _____

Teacher's Notes

Activities 9–12

Ordering and Comparing

Materials
- Numbered Hundred Board
- Worksheet 9: Before and After
- Worksheet 10: Missing Numbers
- Worksheet 11: More Missing Numbers
- Worksheet 12: What's Missing?

The activities on Worksheets 9–12 allow pupils to practice ordering and comparing numbers. Before beginning Worksheet 12, check that pupils understand the column patterns. Count each column for practice.

1	2	3	4	5	6	7	8	9	10
11	12	13	14	15	16	17	18	19	20
21	22	23	24	25	26	27	28	29	30
31	32	33	34	35	36	37	38	39	40
41	42	43	44	45	46	47	48	49	50
51	52	53	54	55	56	57	58	59	60
61	62	63	64	65	66	67	68	69	70
71	72	73	74	75	76	77	78	79	80
81	82	83	84	85	86	87	88	89	90
91	92	93	94	95	96	97	98	99	100

Activity 13

The Calendar

Materials
- Numbered Hundred Board
- Worksheet 13: The Calendar

Lead a discussion in which the pupils compare and contrast a one-month calendar to a hundred board. In your discussion, include the following questions:

- How many rows on a hundred board? On a calendar?
- How many squares on a calendar? On a hundred board?
- How many numbers on a calendar? On a hundred board?

Hand out copies of Worksheet 13, and ask pupils to solve the problems.

Activity 14

Making a Calendar

Materials
- Worksheet 14: This Month

Complete the calendar using information from the current month. Use this information to answer the questions.

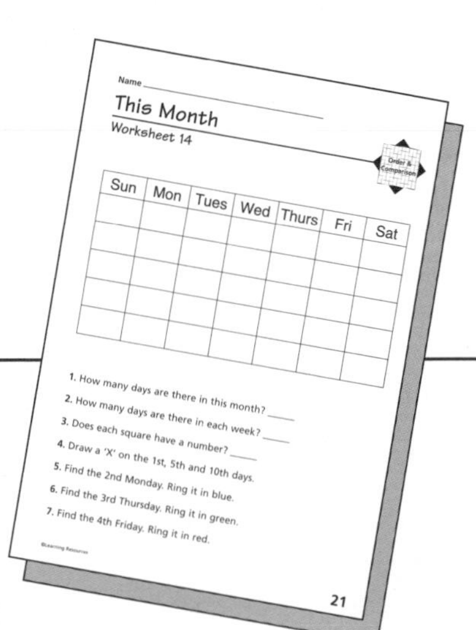
Name

This Month

Worksheet 14

Order & Comparison

Sun	Mon	Tues	Wed	Thurs	Fri	Sat

1. How many days are there in this month? ____
2. How many days are there in each week? ____
3. Does each square have a number? ____
4. Draw a 'X' on the 1st, 5th and 10th days.
5. Find the 2nd Monday. Ring it in blue.
6. Find the 3rd Thursday. Ring it in green.
7. Find the 4th Friday. Ring it in red.

21

Activity 15

High/Low Game

Materials • Numbered Hundred Board
• Transparent counters

Ask each pupil to cover the numbers on the Numbered Hundred Board as they are called out by the teacher. Call two, three, or four numbers at a time. Encourage pupils to place three counters on the highest number and one on the lowest lowest number in each group.

1	2	3	4	5	6	7	8	9	10
11	12	13	14	15	16	17	18	19	20
21	22	23	24	25	26	27	28	29	30
31	32	33	34	35	36	37	38	39	40
41	42	43	44	45	46	47	48	49	50
51	52	53	54	55	56	57	58	59	60
61	62	63	64	65	66	67	68	69	70
71	72	73	74	75	76	77	78	79	80
81	82	83	84	85	86	87	88	89	90
91	92	93	94	95	96	97	98	99	100

Activity 16

Ordering Tiles

Materials • Number tiles

Randomly distribute six to ten number tiles to each pupil. Instruct children to place the numbers in order from lowest to highest, or highest to lowest. Keep the activity going by trading tiles with neighbors.

Activity 17

Scrambled Tiles

Materials • Number tiles

Organize the class into cooperative groups. Give each group a pile of number tiles. Instruct children to place the tiles in order, by evens or odds, or in an original sequence. Groups may wish to have races with each other.

Teacher's Notes

Activity 18

Number Scramble

Materials
- Blank Hundred Board
- Number tiles

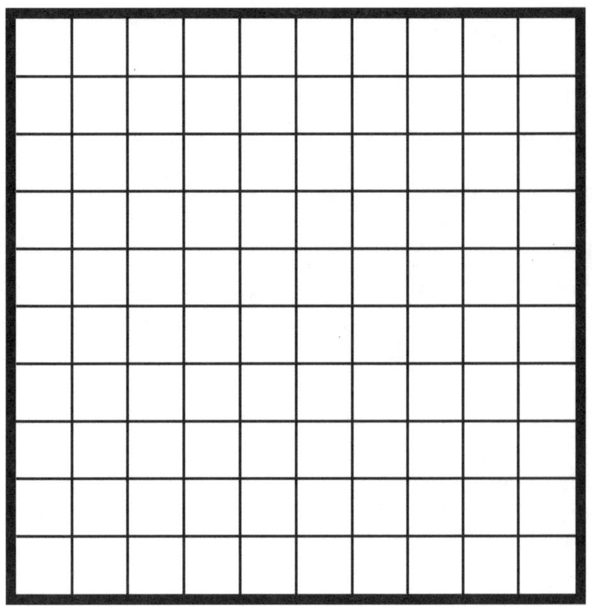

This is a game for two or three players. Randomly select ten number tiles and place them on the Blank Hundred Board in the correct places. Divide the remaining tiles equally between the players. Each player may place a tile before, after, above, or below a number already on the board. The numbers must be in order from lowest to highest or a predetermined order. Players may place one tile per turn. Players who cannot place a tile must pass. The game continues until one player has placed all of their tiles.

Activity 19

Keep Picking

Materials
- Blank Hundred Board
- Number tiles
- Paper bag

This is a number-order game for two to four players. Each player randomly picks ten number tiles and places them face-up. Place the remaining tiles in a paper bag. Players take turns placing their tiles in sequence on the Blank Hundred Board, either horizontally or vertically.

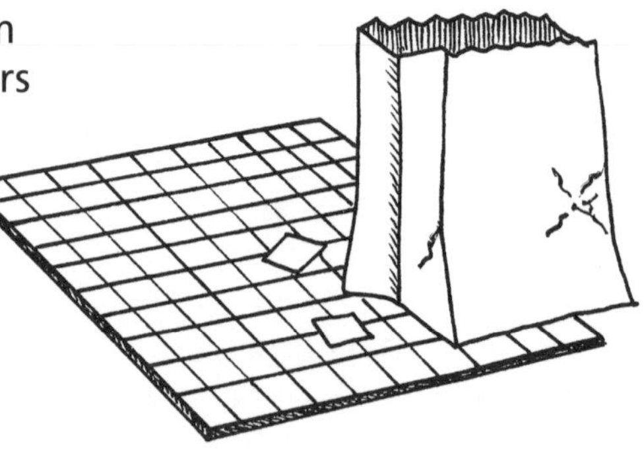

Players may place as many tiles as possible each turn by adding to the numbers already on the board, or by beginning a new sequence. (Players must have two or more numbers to start a new sequence.)

Players who cannot place any of their tiles must draw additional tiles from the bag until a play can be made. The first player to use all of his or her tiles wins.

Activity 20

Greater Than and Less Than

Materials • Counters

For an introductory activity, distribute a random number of counters to each pupil. Working in pairs, each pupil counts their number of counters. Which partner has a greater number? Go around the room asking pupils to tell their number sentences (Earl has 12 counters. Joyce has 8 counters. Earl has a greater number of counters than Joyce.)

Activity 21

Which Has More?

Materials • Blank Hundred Board
• Counters

Give each player twenty counters (ten each of two colors). Using the Blank Hundred Board, ask pupils to place five counters of one color in the first row. In the second row, line up seven counters of the second color. It will be visible which row has more counters. Explain that the relationship between the rows can be written using the symbols > (greater than) and < (less than). Continue with more examples.

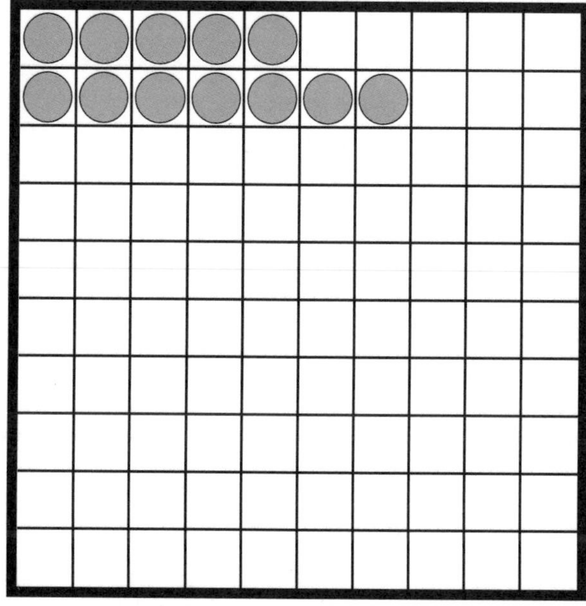

5 is less than 7 **$5 < 7$**
7 is greater than 5 **$7 > 5$**

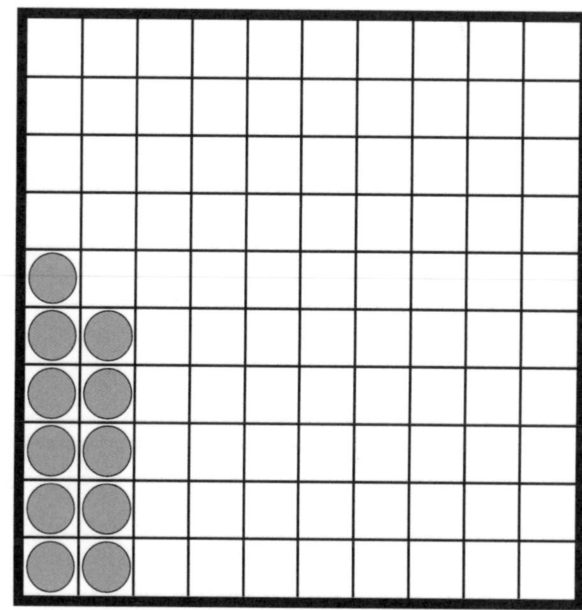

6 is greater than 5 **$6 > 5$**
5 is less than 6 **$5 < 6$**

Teacher's Notes

Activity 22

I've Got It Covered!

Materials
- Numbered Hundred Board
- Transparent counters

This game explores numbers "greater than" and "less than." Give each pupil a hundred board and several transparent counters. Call out two numbers to cover. Children should cover the greater number with two counters. Encourage verbal response from students. Example: When 30 and 12 are called out, pupils cover the numbers and then respond with either "30 is greater than 12" or "12 is less than 30."

1	2	3	4	5	6	7	8	9	10
11	12	13	14	15	16	17	18	19	20
21	22	23	24	25	26	27	28	29	30
31	32	33	34	35	36	37	38	39	40
41	42	43	44	45	46	47	48	49	50
51	52	53	54	55	56	57	58	59	60
61	62	63	64	65	66	67	68	69	70
71	72	73	74	75	76	77	78	79	80
81	82	83	84	85	86	87	88	89	90
91	92	93	94	95	96	97	98	99	100

Activities 23-24

Greater Than or Less Than

Materials
- Worksheet 23: Greater Than or Less Than
- Worksheet 24: Do You Know?

The activities on Worksheets 23 and 24 allow pupils to practice comparing number pairs. Before handing out copies of Worksheet 23, lead a "greater than" and "less than" discussion relating to the classroom. Is the number of girls greater than or less than the number of boys in our class? Is the number of tables greater than or less than the number of chairs?

Together, work through the first number sentence on Worksheet 23, then let children complete the problems independently. Worksheet 24 provides further practice in ordering numbers.

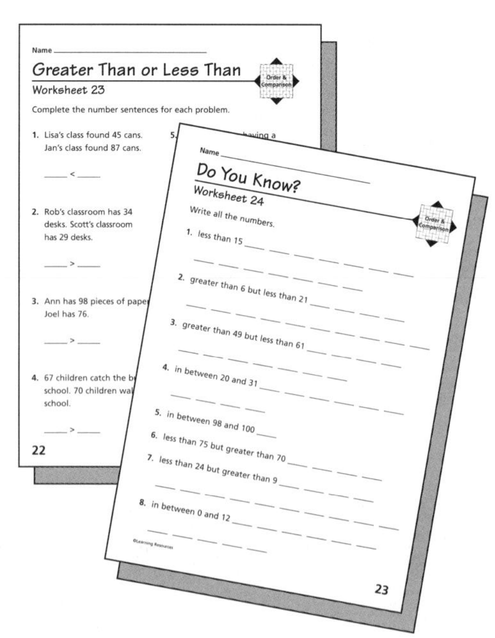
Name ____
Greater Than or Less Than
Worksheet 23
Complete the number sentences for each problem.
1. Lisa's class found 45 cans. Jan's class found 87 cans.
___ < ___
2. Rob's classroom has 34 desks. Scott's classroom has 29 desks.
___ > ___
3. Ann has 98 pieces of pape... Joel has 76.
___ > ___
4. 67 children catch the b... school. 70 children wal... school.
___ > ___
22

Name ____
Do You Know?
Worksheet 24
Write all the numbers.
1. less than 15
2. greater than 6 but less than 21
3. greater than 49 but less than 61
4. in between 20 and 31
5. in between 98 and 100
6. less than 75 but greater than 70
7. less than 24 but greater than 9
8. in between 0 and 12
23

Teacher's Notes

Activity 25

Ordinals: First-Fifth

Materials
- Crayons: blue, red, green, purple, orange, brown, black
- Pencils
- Worksheet 25: Ordinal Numbers

Hand out copies of Worksheet 25 with sets of colored crayons and pencils. Ask pupils to follow these directions:

With a *blue* crayon, write in the first number on Worksheet 25's hundred board.
With a *red* crayon, write in the second number.
With a *green* crayon, write in the third number.
With a *purple* crayon, write in the fourth number.
With an *orange* crayon, write in the fifth number.
With a pencil, shade the remaining numbers in the first row

Fill in the second row of numbers with a *brown* crayon.
Fill in the third row of numbers with a *black* crayon.
Fill in the fourth row of numbers with a *red* crayon.
Fill in the fifth row of numbers with a *blue* crayon.

1st row										
2nd row										
3rd row										
4th row										
5th row										
	51	52	53	54	55	56	57	58	59	60
	61	62	63	64	65	66	67	68	69	70
	71	72	73	74	75	76	77	78	79	80
	81	82	83	84	85	86	87	88	89	90
	91	92	93	94	95	96	97	98	99	100

Name ________________________________

Before and After

Order & Comparison

Worksheet 9

Put a counter on the board to cover the number that comes after each of these numbers.

1. 7 **2.** 13 **3.** 0 **4.** 46

Put a counter on the board to cover the number that comes before each of these numbers.

5. 6 **6.** 53 **7.** 100 **8.** 22

Write the numbers that come before and after these numbers.

9. ______ , 7, ______

10. ______ , 14, ______

11. ______ , 20, ______

12. ______ , 65, ______

13. ______ , 78, ______

14. ______ , 83, ______

Find these numbers. Try to make a pattern.

15. Write the three numbers that come before 27.

_____ _____ _____

16. Write the five numbers that come after 39.

_____ _____ _____ _____ _____

Name ______________________________

Missing Numbers

Worksheet 10

Order & Comparison

Write the missing numbers in each row.

1		3		5			8		10
	12			15		17			20
		23			26			29	
31			34			37			
	42			45			48		50
		53			56			59	

Name ________________________________

More Missing Numbers

Worksheet 11

Order & Comparison

Fill in the missing numbers to complete the board.

1		3		5		7			10
	12				16				
			24				28		
				35				39	
41						47			
	52								60
		63						69	
			74			77			
		83			86				
							98		100

Name ______________________________

What's Missing?

Order & Comparison

Worksheet 12

Write the missing numbers.

		10
	19	

	3	
	23	
		34

27		
	58	

		79
	98	

		57
	66	

72	

Name ______________________________

The Calendar

Worksheet 13

Sun.	Mon.	Tues.	Wed.	Thurs.	Fri.	Sat.
	1	2	3	4	5	6
7	8	9	10	11	12	13
14	15	16	17	18	19	20
21	22	23	24	25	26	27
28	29	30	31			

How many rows are...

1. on a hundred board? _____

2. on the calendar? _____

How many numbers are there altogether...

3. on a hundred board? _____

4. on the calendar? _____

How many numbers are in each row...

5. on a hundred board? _____

6. on the calendar? _____

What is the first number...

7. on a hundred board? _____

8. on the calendar? _____

Name ______________________________

This Month

Worksheet 14

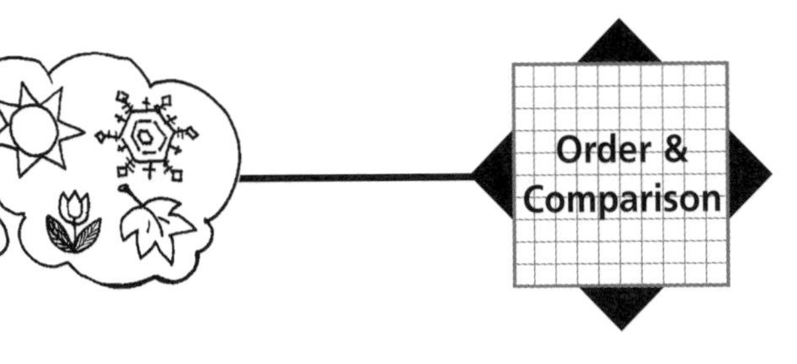

Order & Comparison

Sun.	Mon.	Tues.	Wed.	Thurs.	Fri.	Sat.

Use the blank calendar page to answer the following questions.

1. Count the number of days this month. _____

2. How many days are there in each week? _____

3. Do all squares have numbers? _____

4. Draw an 'X' on the 1st, 5th, and 10th.

5. Find the 2nd Monday. Circle it in blue.

6. Find the 3rd Thursday. Circle it in green.

7. Find the 4th Friday. Circle it in red.

Name ______________________________

Greater Than or Less Than

Order & Comparison

Worksheet 23

Complete the number sentences for each problem.

1. Lisa's class found 45 cans. Jan's class found 87 cans.

2. Rob's classroom has 34 desks. Scott's classroom has 29 desks.

3. Ann has 98 pieces of paper. Joel has 76.

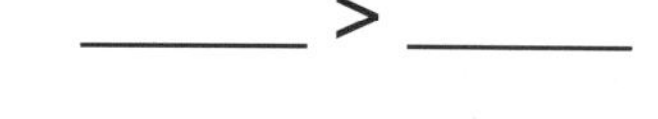

4. 67 children catch the bus to school. 70 children walk to school.

5. 12 children are eating a hot lunch. 16 children brought sack lunches.

______ < ______

6. 20 children have finished their work. 8 are still working.

______ > ______

7. 58 children are at recess. 86 children are working in their classrooms.

______ > ______

Name ____________________

Do You Know?

Order & Comparison

Worksheet 24

Write all the numbers in the correct order.

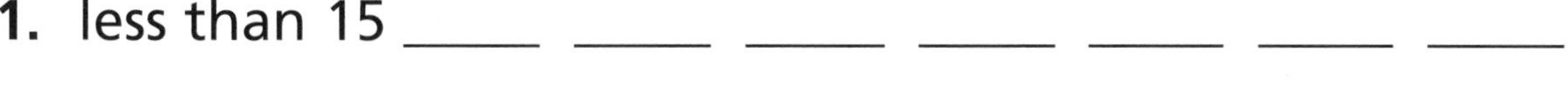

1. less than 15 _____ _____ _____ _____ _____ _____ _____

_____ _____ _____ _____ _____ _____ _____

2. greater than 6 but less than 21 _____ _____ _____ _____

_____ _____ _____ _____ _____ _____ _____ _____ _____ _____

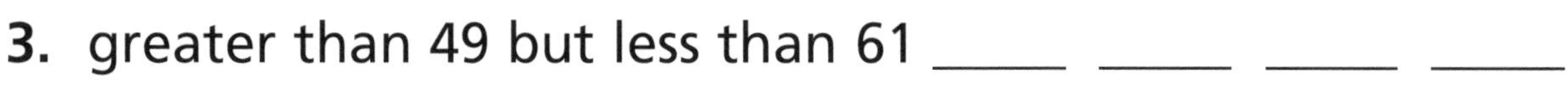

3. greater than 49 but less than 61 _____ _____ _____ _____

_____ _____ _____ _____ _____ _____ _____

4. in between 20 and 31 _____ _____ _____ _____ _____ _____

_____ _____ _____ _____

5. in between 98 and 100 _____

6. less than 75 but greater than 70 _____ _____ _____ _____

7. less than 24 but greater than 9 _____ _____ _____ _____

_____ _____ _____ _____ _____ _____ _____ _____ _____ _____

8. in between 0 and 12 _____ _____ _____ _____ _____ _____

_____ _____ _____ _____ _____

Name ______________________________

Ordinal Numbers

Worksheet 25

Follow your teacher's directions to complete the board below.

51	52	53	54	55	56	57	58	59	60
61	62	63	64	65	66	67	68	69	70
71	72	73	74	75	76	77	78	79	80
81	82	83	84	85	86	87	88	89	90
91	92	93	94	95	96	97	98	99	100

Teacher's Notes

Activity 26

Tens and Ones

Materials
- Numbered Hundred Board
- Transparent counters
- Worksheet 26: How Many Tens and Ones?

1	2	3	4	5	6	7	8	9	10
11	12	13	14	15	16	17	18	19	20
21	22	23	24	25	26	27	28	29	30
31	32	33	34	35	36	37	38	39	40
41	42	43	44	45	46	47	48	49	50
51	52	53	54	55	56	57	58	59	60
61	62	63	64	65	66	67	68	69	70
71	72	73	74	75	76	77	78	79	80
81	82	83	84	85	86	87	88	89	90
91	92	93	94	95	96	97	98	99	100

Explain that a two-digit number has two places, the "tens" place and the "ones" place. State place values (one ten, two ones)(12) aloud, asking children to place counters on that square.

State a number and ask children questions about the number. Encourage children to use the appropriate vocabulary words.

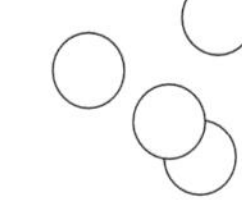

- Which number is in the "ones" place?
- In the "tens" place?
- How many tens in the number?
- How many ones in the number?

Worksheet 26 may be used for additional practice.

Activity 27

Follow Me

Materials
- Numbered Hundred Board
- Pencils
- Blue, red, and green crayons
- Worksheet 27: Which Number Am I?

Children can practice identifying the position of ones and tens by following these directions:

- Circle in *blue* all numbers with 5 in the ones place.
- Circle in *red* all numbers with 0 in the ones place.
- Circle in *green* all numbers with 1 in the ones place.
- Shade in all the numbers that have the same tens and ones numbers (ex.: 22).

Before handing out Worksheet 27, practice working with tens and ones on the chalkboard.

Teacher's Notes

Activity 28

Tens and Ones

Materials
- Numbered Hundred Board
- Opaque counters

After children have completed the following instructions, read the answers as a class.

Cover all numbers with a 2 in the ones place.
Cover all numbers with a 4 in the ones place.
Cover all numbers that have the same number in the ones and tens places.
Cover all numbers with a 5 in the tens place.
Cover all numbers with a 7 in the tens place.
Cover all numbers with a 0 in the ones place.

1	2	3	4	5	6	7	8	9	10
●	12	13	14	15	16	17	18	19	20
21	●	23	24	25	26	27	28	29	30
31	32	●	34	35	36	37	38	39	40
41	42	43	●	45	46	47	48	49	50
51	52	53	54	●	56	57	58	59	60
61	62	63	64	65	●	67	68	69	70
71	72	73	74	75	76	●	78	79	80
81	82	83	84	85	86	87	●	89	90
91	92	93	94	95	96	97	98	●	100

Activity 29

Around the World

Form a circle so children can see each other as they play Around the World. Children will take turns one at a time. The teacher begins by making a place value statement (six tens and six ones). One child will answer with the correct number (66) and pose a place value statement to his neighbor. Play continues "around the world" until all children have had a turn.

For example:.

Teacher:	Six tens and six ones.
Pupil A:	66. Eight tens and three ones.
Pupil B:	83. Seven tens and seven ones.
Pupil C:	77. Four tens and no ones.
Pupil D:	40. One ten and nine ones.

Activity 30

Tens and Ones Bingo

Materials
- Numbered Hundred Board
- Transparent counters
- Number tiles
- Paper bag

Tens and ones place values are called out and children must decide which number to cover. For example, 43 is called as four tens, three ones; 50 as five tens, no ones. The caller may be the child whose birthday is closest to the current date. The caller chooses number tiles from a bag and calls the place value. Whoever completes a row horizontally or vertically wins, and becomes the caller for the next game.

1	2	3	4	5	6	7	8	9	10
11	12	13	14	15	16	17	18	19	20
21	22	23	24	25	26	27	28	29	30
31	32	33	34	35	36	37	38	39	40
41	42	43	44	45	46	47	48	49	50
51	52	53	54	55	56	57	58	59	60
61	62	63	64	65	66	67	68	69	70
71	72	73	74	75	76	77	78	79	80
81	82	83	84	85	86	87	88	89	90
91	92	93	94	95	96	97	98	99	100

26
35
62

Activity 31-32

Identifying Tens and Ones

Materials
- Numbered Hundred Board
- Worksheets 31: Lowest or Highest?
- Worksheets 32: Number Facts

The activities on Worksheets 31 and 32 allow pupils to practice identifying ones and tens places. Before distributing the worksheets, review the ones and tens positions in each column and row on the hundred board.

Activity 33

Number Sorts

Materials
- Blank Hundred Board
- Number tiles 1-100
- Worksheet 33: Number Sorts

The activities on Worksheet 33 are suggested for two to four students. Children take turns completing the activities, replacing the tiles after each question.

Children begin each activity by placing the number tiles 1-100 in order on the Blank Hundred Board.

Name ______________________________

How Many Tens and Ones?

Place Value

Worksheet 26

Write the numbers that have tens only (zero ones).

____ ____ ____ ____ ____ ____ ____ ____ ____ ____

How many tens and ones are in each number below?

Number	Tens	Ones
36	3	6
23		
12		
61		
4		
30		

Name ______________________________

Which Number Am I?

Worksheet 27

Write the following numbers.

1. 6 tens and 6 ones _____
2. 3 tens and 7 ones _____
3. 7 tens and 3 ones _____
4. 1 ten and 0 ones _____
5. the highest number with 8 ones _____
6. the lowest number with 2 tens _____

Write all the numbers:

7. with 3 ones

_____ _____ _____ _____ _____

_____ _____ _____ _____ _____

8. with 3 tens

_____ _____ _____ _____ _____

_____ _____ _____ _____ _____

Name ______________________________

Lowest or Highest?

Worksheet 31

Place Value

Using the Blank Hundred Board, write the numbers that are described below.

1. the lowest two-digit number _____

2. the highest two-digit number _____

3. has a 5 in both the tens place and the ones place _____

4. has a 4 in the tens place and a 6 in the ones place _____

5. has the same number in both the tens place and the ones place

_____ , _____ , _____ , _____ , _____ , _____ , _____ , _____ , _____

6. all two-digit numbers with 0 ones

_____ , _____ , _____ , _____ , _____ , _____ , _____ , _____ , _____

Write the highest number:

7. with 7 tens _____

8. with 7 ones _____

Write the lowest number:

9. with 0 ones _____

10. that has three digits _____

Name ______________________________

Number Facts

Place Value

Worksheet 32

Use the Numbered Hundred Board to find the following numbers.

1. the highest 2-digit number on the hundred board _____

2. the lowest 2-digit number on the hundred board _____

3. the 3rd number in the 3rd row _____

4. the 4th number in the 7th row _____

5. greater than 74 but less than 76 _____

6. less than 68 but greater than 66 _____

7. greater than 43 but less than 45 _____

8. less than 86 but greater than 84 _____

9. the highest one-digit number _____

10. the 5th number in the 2nd row _____

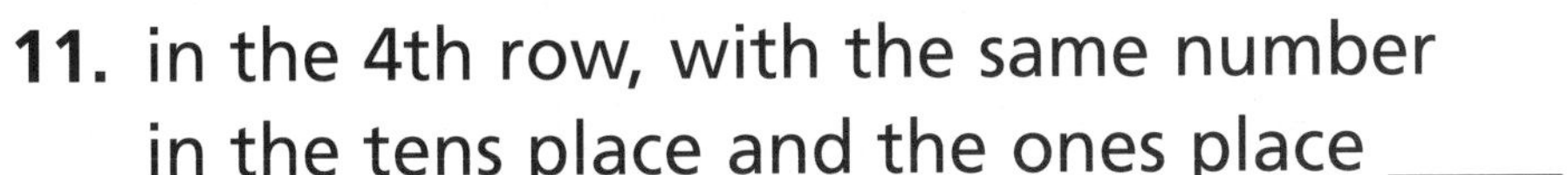

11. in the 4th row, with the same number in the tens place and the ones place _____

12. the lowest number with both digits the same _____

Name ______________________________

Number Sorts

Place Value

Worksheet 33

Place the number tiles in order from 1-100. Find these numbers:

1. the same in both the tens and the ones place

_____ _____ _____ _____ _____ _____ _____ _____ _____

2. greater than 49 and less than 60

_____ _____ _____ _____ _____ _____ _____ _____ _____ _____

3. with digits that add up to 5

_____ _____ _____ _____ _____

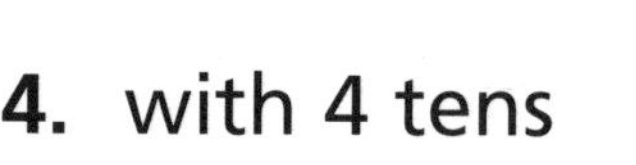

4. with 4 tens

_____ _____ _____ _____ _____ _____ _____ _____ _____ _____

5. with 8 ones

_____ _____ _____ _____ _____ _____ _____ _____ _____ _____

6. with digits that add up to 10

_____ _____ _____ _____ _____ _____ _____ _____ _____

7. between 89 and 100

_____ _____ _____ _____ _____ _____ _____ _____ _____ _____

Teacher's Notes

Activity 34

Counting in Tens

Materials
- Blank Hundred Board
- Number tiles
- Transparent counters, ten for each child

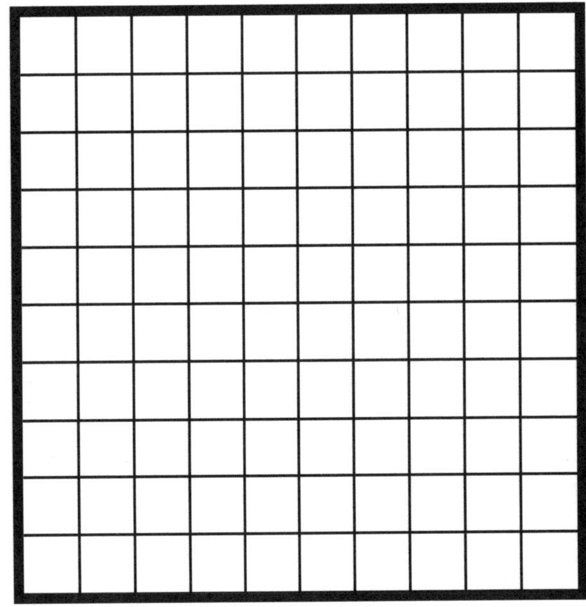

Ask pupils to place number tiles 1-100 on the correct squares on the Blank Hundred Board. Instruct them to place a transparent counter on the number 10. Continue counting by tens, placing counters on each multiple of ten.

Activity 35

Counting in Fives

Materials
- Blank Hundred Board
- Number tiles
- Transparent counters, twenty for each child
- Worksheet 35: Patterns of Ten and Five

Ask pupils to place number tiles 1-100 on the correct squares on the Blank Hundred Board. Instruct them to place a transparent counter on the number 5. Continue counting by fives, placing counters on each multiple of five.

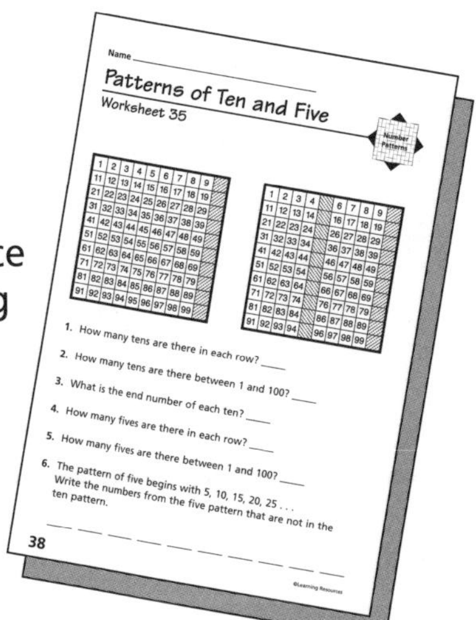

Name

Patterns of Ten and Five

Worksheet 35

Number Patterns

1	2	3	4	5	6	7	8	9	
11	12	13	14	15	16	17	18	19	
21	22	23	24	25	26	27	28	29	
31	32	33	34	35	36	37	38	39	
41	42	43	44	45	46	47	48	49	
51	52	53	54	55	56	57	58	59	
61	62	63	64	65	66	67	68	69	
71	72	73	74	75	76	77	78	79	
81	82	83	84	85	86	87	88	89	
91	92	93	94	95	96	97	98	99	

1	2	3	4		6	7	8	9	
11	12	13	14		16	17	18	19	
21	22	23	24		26	27	28	29	
31	32	33	34		36	37	38	39	
41	42	43	44		46	47	48	49	
51	52	53	54		56	57	58	59	
61	62	63	64		66	67	68	69	
71	72	73	74		76	77	78	79	
81	82	83	84		86	87	88	89	
91	92	93	94		96	97	98	99	

1. How many tens are there in each row? ____
2. How many tens are there between 1 and 100? ____
3. What is the end number of each ten? ____
4. How many fives are there in each row? ____
5. How many fives are there between 1 and 100? ____
6. The pattern of five begins with 5, 10, 15, 20, 25 . . . Write the numbers from the five pattern that are not in the ten pattern.

____ ____ ____ ____ ____ ____ ____ ____ ____ ____

38

©Learning Resources

Worksheet 35 provides further practice for counting in multiples of five and ten.

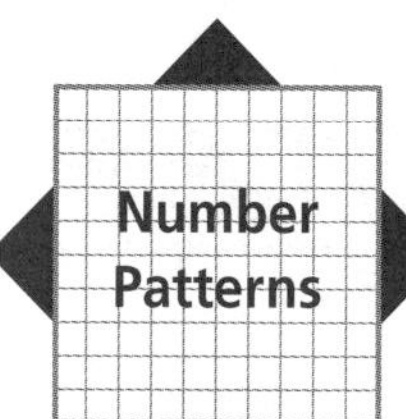

Teacher's Notes

Activity 36

A Pattern of Three

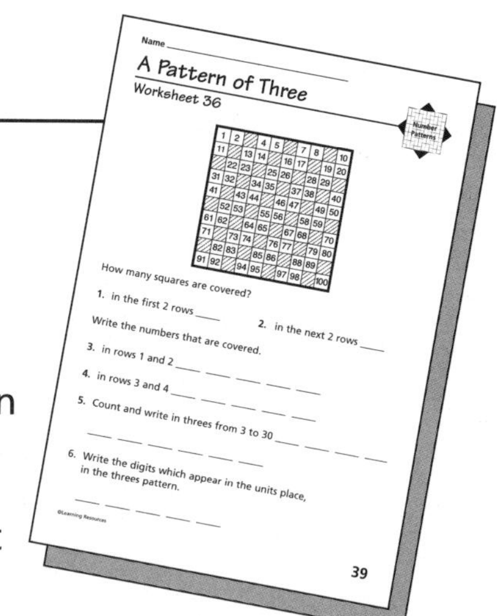
Name

A Pattern of Three

Worksheet 36

How many squares are covered?

1. in the first 2 rows ___
2. in the next 2 rows ___

Write the numbers that are covered.

3. in rows 1 and 2 ___ ___ ___ ___ ___ ___
4. in rows 3 and 4 ___ ___ ___ ___ ___ ___
5. Count and write in threes from 3 to 30 ___ ___ ___ ___ ___ ___ ___ ___ ___ ___
6. Write the digits which appear in the units place, in the threes pattern.

©Learning Resources

39

Materials ● Numbered Hundred Board
● Worksheet 36: A Pattern of Three

Distribute Numbered Hundred Board worksheets. Ask children to follow these directions:
•Shade square number 3. •Skip the next two squares and shade number 6. •Skip two more squares and shade the next number. •Continue skipping two squares and shading until you reach 99.

When finished, ask if anyone sees a pattern. Introduce the term *diagonal* if necessary. Use Worksheet 36 for reinforcement.

1	2	3	4	5	6	7	8	9	10
11	12	13	14	15	16	17	18	19	20
21	22	23	24	25	26	27	28	29	30
31	32	33	34	35	36	37	38	39	40
41	42	43	44	45	46	47	48	49	50
51	52	53	54	55	56	57	58	59	60
61	62	63	64	65	66	67	68	69	70
71	72	73	74	75	76	77	78	79	80
81	82	83	84	85	86	87	88	89	90
91	92	93	94	95	96	97	98	99	100

Activity 37

Twos and Threes

Materials ● Numbered Hundred Board
● Transparent counters, 33 for each group

Working in small groups, ask children to cover square number 2. Then, skip a square and cover square 4. Continue this pattern until you reach 100. When finished, challenge children to count by threes, placing a counter over all multiples of three. Ask children to find the numbers that are multiples of both two and three (6, 12, 18, 24, 30, 36, 42, 48, 54, 60, 66, 72, 78, 84, 90, 96).

Activity 38

A Pattern of Four

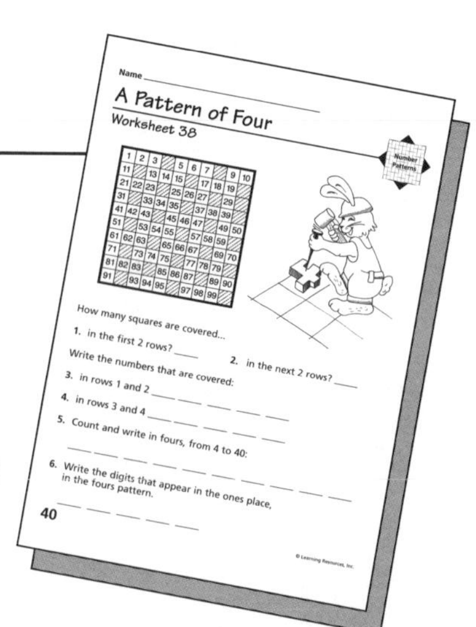
Name

A Pattern of Four

Worksheet 38

How many squares are covered...

1. in the first 2 rows? ___
2. in the next 2 rows? ___

Write the numbers that are covered:

3. in rows 1 and 2 ___ ___ ___ ___ ___
4. in rows 3 and 4 ___ ___ ___ ___ ___
5. Count and write in fours, from 4 to 40: ___ ___ ___ ___ ___ ___ ___ ___ ___ ___
6. Write the digits that appear in the ones place, in the fours pattern.

40

© Learning Resources, Inc.

Materials ● Numbered Hundred Board
● Worksheet 38: A Pattern of Four

Ask pupils to shade the number 4. Then skip the next three squares and shade number 8. Continue skipping three squares and shading the fourth, until the board is shaded with multiples of four. Ask children to describe the pattern that results. Use Worksheet 38 for more practice.

Teacher's Notes

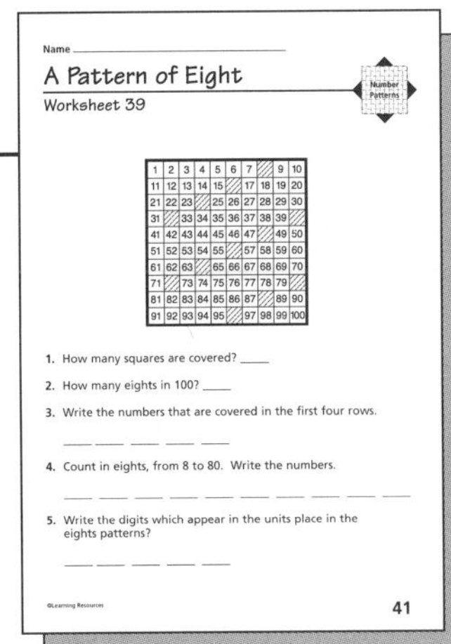
Name

A Pattern of Eight

Worksheet 39

1. How many squares are covered? ____
2. How many eights in 100? ____
3. Write the numbers that are covered in the first four rows.
4. Count in eights, from 8 to 80. Write the numbers.
5. Write the digits which appear in the units place in the eights patterns?

©Learning Resources 41

Activity 39

A Pattern of Eight

Materials
- Numbered Hundred Board
- Worksheet 39: A Pattern of Eight

Ask pupils to cover square number 8. Count eight more and shade square 16. Continue shading every eighth square, watching the pattern grow each square is shaded. Make sure every square fits the pattern before you shade. Continue until square 96 and discuss the pattern.

Use Worksheet 39 for more practice.

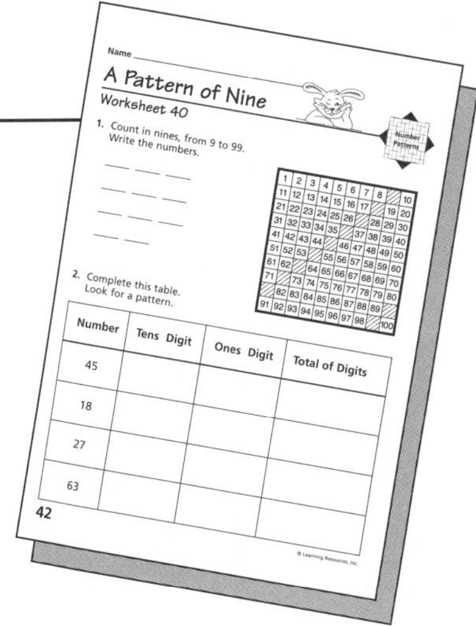
Name

A Pattern of Nine

Worksheet 40

1. Count in nines, from 9 to 99. Write the numbers.
2. Complete this table. Look for a pattern.

Number	Tens Digit	Ones Digit	Total of Digits
45			
18			
27			
63			

42 © Learning Resources, Inc.

Activity 40

A Pattern of Nine

Materials
- Numbered Hundred Board
- Worksheet 40: A Pattern of Nine

Ask pupils to shade square number 9 on the Numbered Hundred Board. Count nine more and shade square 18. Ask children if they can predict a pattern. Continue shading every ninth square. (Number 81 is not the last number.) Continue until you reach square 99.

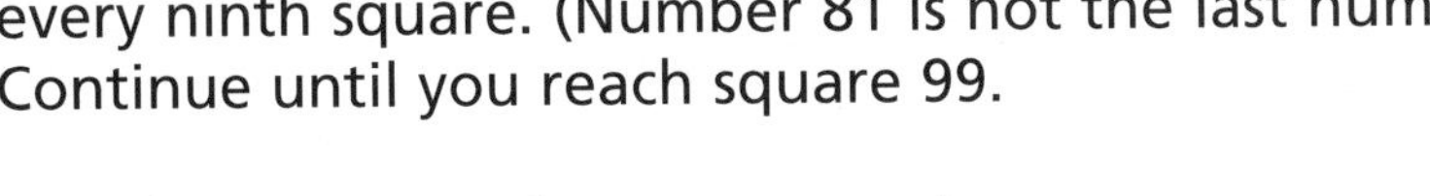

Use Worksheet 40 for more practice.

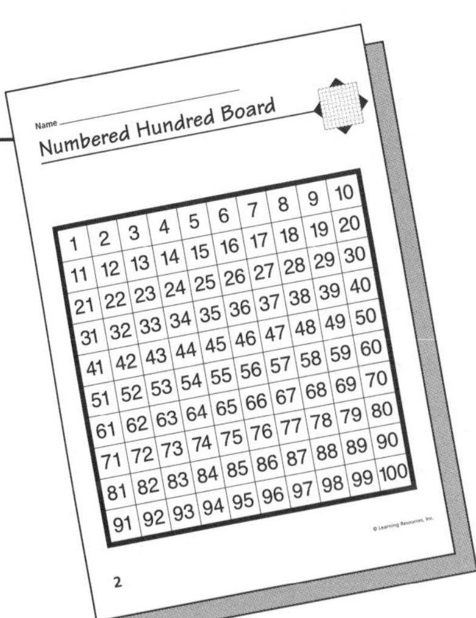
Name

Numbered Hundred Board

1	2	3	4	5	6	7	8	9	10
11	12	13	14	15	16	17	18	19	20
21	22	23	24	25	26	27	28	29	30
31	32	33	34	35	36	37	38	39	40
41	42	43	44	45	46	47	48	49	50
51	52	53	54	55	56	57	58	59	60
61	62	63	64	65	66	67	68	69	70
71	72	73	74	75	76	77	78	79	80
81	82	83	84	85	86	87	88	89	90
91	92	93	94	95	96	97	98	99	100

2 © Learning Resources, Inc.

Activity 41

Threes and Nines

Materials
- Numbered Hundred Board
- Transparent counters, 11 for each group

Organize children into small groups. Ask them to cover multiples of threes on their hundred boards. Next, instruct students to cover multiples of nine. Find numbers that are multiples of three and nine. How are the patterns related? (Every third number in the threes pattern is a member of the nines pattern.)

Teacher's Notes

Activity 42

Even and Odd

Materials
- Numbered Hundred Board
- Transparent counters, fifty for each pupil
- Worksheet 42: Even and Odd Patterns

The following activities reinforce skills using even and odd numbers. Begin by reviewing the proper way to identify and label intervals on the hundred boards.

Pupils may work independently or with partners for the following activities. Instruct them to cover every other number, starting with 2. When finished, ask children to identify the numbers in the ones place (0, 2, 4, 6, 8). Explain that these are even numbers, and that all even numbers end in a 0, 2, 4, 6, or 8, and can be divided evenly.

1	(2)	3	(4)	5	(6)	7	(8)	9	(10)
11	(12)	13	(14)	15	(16)	17	(18)	19	(20)
21	(22)	23	(24)	25	(26)	27	(28)	29	(30)
31	(32)	33	(34)	35	(36)	37	(38)	39	(40)
41	(42)	43	(44)	45	(46)	47	(48)	49	(50)
51	(52)	53	(54)	55	(56)	57	(58)	59	(60)
61	(62)	63	(64)	65	(66)	67	(68)	69	(70)
71	(72)	73	(74)	75	(76)	77	(78)	79	(80)
81	(82)	83	(84)	85	(86)	87	(88)	89	(90)
91	(92)	93	(94)	95	(96)	97	(98)	99	(100)

Ask:

How many covered squares are in each row?
How many even numbers are there between 1 and 100?
What is the second even number?
What is the seventh even number?
What is the twelfth even number?

Follow the same procedure for the numbers that are not covered. Cover every other number, starting with 1. When finished, ask children to identify the numbers in the ones place. (1, 3, 5, 7, 9). Explain that these are odd numbers; they cannot be divided evenly.

Ask:

How many covered squares are in each row?
How many odd numbers are there between 1 and 100?
What is the second odd number?
What is the seventh odd number?
What is the twelfth odd number?

Teacher's Notes

Activity 43

Practice with Odds and Evens

Materials
- Red and blue crayons
- Numbered Hundred Board
- Worksheet 43: Even and Odd

To practice the concept of even and odd numbers, distribute Numbered Hundred Board worksheets. Ask students to circle all even numbers in blue and all odd numbers in red. Are any numbers leftover? (No, every number is either even or odd.)

The questions on Worksheet 43 provide more work with odd and even numbers. Pupils can use the hundred board to help them answer questions.

Activity 44

Odd/Even Bingo

Materials
- Numbered Hundred Board
- Bag of counters, in two colors
- Die numbered 1-6

This is a game for two players. Use spaces 1-50 on the hundred board for the game. Give one player blue counters and the other player red counters. Players take turns rolling the die. If an even number is rolled, the player may cover any even number that is not already covered. If a player rolls an odd number, an odd number from 1-50 may be covered. The first player to cover five spaces in a row, in any direction, wins.

1	2	3	4	5	6	7	8	9	10
11	12	13	14	15	16	17	18	19	20
21	22	23	24	25	26	27	28	29	30
31	32	33	34	35	36	37	38	39	40
41	42	43	44	45	46	47	48	49	50
51	52	53	54	55	56	57	58	59	60
61	62	63	64	65	66	67	68	69	70
71	72	73	74	75	76	77	78	79	80
81	82	83	84	85	86	87	88	89	90
91	92	93	94	95	96	97	98	99	100

Name ____________________________________

Patterns of Ten and Five

Worksheet 35

1	2	3	4	5	6	7	8	9	
11	12	13	14	15	16	17	18	19	
21	22	23	24	25	26	27	28	29	
31	32	33	34	35	36	37	38	39	
41	42	43	44	45	46	47	48	49	
51	52	53	54	55	56	57	58	59	
61	62	63	64	65	66	67	68	69	
71	72	73	74	75	76	77	78	79	
81	82	83	84	85	86	87	88	89	
91	92	93	94	95	96	97	98	99	

1	2	3	4		6	7	8	9	
11	12	13	14		16	17	18	19	
21	22	23	24		26	27	28	29	
31	32	33	34		36	37	38	39	
41	42	43	44		46	47	48	49	
51	52	53	54		56	57	58	59	
61	62	63	64		66	67	68	69	
71	72	73	74		76	77	78	79	
81	82	83	84		86	87	88	89	
91	92	93	94		96	97	98	99	

1. How many tens are there in each row? _____

2. How many tens are there between 1 and 100? _____

3. What number is in the ones place of each multiple of ten? _____

4. How many fives are in each row? _____

5. How many fives fall between 1 and 100? _____

6. The pattern of five begins with 5, 10, 15, 20, 25 . . .
   Write the numbers from the five pattern that are not in the ten pattern.

   _____ _____ _____ _____ _____ _____ _____ _____ _____ _____

Name ____________________

A Pattern of Three

Worksheet 36

1	2		4	5		7	8		10
11		13	14		16	17		19	20
	22	23		25	26		28	29	
31	32		34	35		37	38		40
41		43	44		46	47		49	50
	52	53		55	56		58	59	
61	62		64	65		67	68		70
71		73	74		76	77		79	80
	82	83		85	86		88	89	
91	92		94	95		97	98		100

How many squares are covered...

1. in the first 2 rows? _____

2. in the next 2 rows? _____

Write the numbers that are covered:

3. in rows 1 and 2 _____ _____ _____ _____ _____ _____

4. in rows 3 and 4 _____ _____ _____ _____ _____ _____ _____

5. Count and write in threes from 3 to 30:

_____ _____ _____ _____ _____ _____ _____ _____ _____ _____

6. Write the digits that appear in the ones place, for all the numbers in the threes pattern.

_____ _____ _____ _____ _____ _____ _____ _____ _____

Name ______________________________

A Pattern of Four

Number Patterns

Worksheet 38

1	2	3		5	6	7		9	10
11		13	14	15		17	18	19	
21	22	23		25	26	27		29	
31		33	34	35		37	38	39	
41	42	43		45	46	47		49	50
51		53	54	55		57	58	59	
61	62	63		65	66	67		69	70
71		73	74	75		77	78	79	
81	82	83		85	86	87		89	90
91		93	94	95		97	98	99	

How many squares are covered...

1. in the first 2 rows? _____

2. in the next 2 rows? _____

Write the numbers that are covered:

3. in rows 1 and 2 _____ _____ _____ _____ _____

4. in rows 3 and 4 _____ _____ _____ _____ _____

5. Count and write in fours, from 4 to 40:

_____ _____ _____ _____ _____ _____ _____ _____ _____ _____

6. Write the digits that appear in the ones place, in the fours pattern.

_____ _____ _____ _____ _____

Name ______________________________

A Pattern of Eight

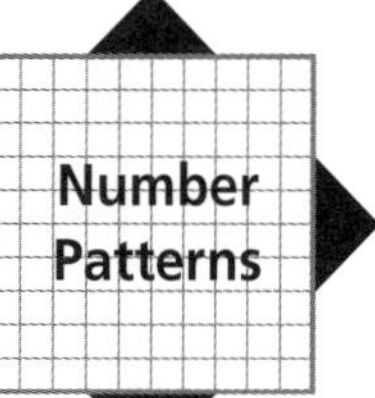

Worksheet 39

1	2	3	4	5	6	7		9	10
11	12	13	14	15		17	18	19	20
21	22	23		25	26	27	28	29	30
31		33	34	35	36	37	38	39	
41	42	43	44	45	46	47		49	50
51	52	53	54	55		57	58	59	60
61	62	63		65	66	67	68	69	70
71		73	74	75	76	77	78	79	
81	82	83	84	85	86	87		89	90
91	92	93	94	95		97	98	99	100

1. How many squares are covered on the hundred board? _____

2. How many eights in 100? _____

3. Write the numbers that are covered in the first four rows.

   _____ _____ _____ _____ _____

4. Count in eights, from 8 to 80. Write the numbers.

   _____ _____ _____ _____ _____ _____ _____ _____ _____ _____

5. Write the digits that appear in the ones place, in the eights patterns.

   _____ _____ _____ _____ _____

Name ____________________________________

A Pattern of Nine

Number Patterns

Worksheet 40

1. Count in nines, from 9 to 99.
   Write the numbers.

   _____ _____ _____

   _____ _____ _____

   _____ _____ _____

   _____ _____

1	2	3	4	5	6	7	8		10
11	12	13	14	15	16	17		19	20
21	22	23	24	25	26		28	29	30
31	32	33	34	35		37	38	39	40
41	42	43	44		46	47	48	49	50
51	52	53		55	56	57	58	59	60
61	62		64	65	66	67	68	69	70
71		73	74	75	76	77	78	79	80
	82	83	84	85	86	87	88	89	
91	92	93	94	95	96	97	98		100

2. Complete this table.
   Look for a pattern.

Number	Tens Digit	Ones Digit	Total of Digits
45			
18			
27			
63			

Name ______________________________

Even and Odd Patterns

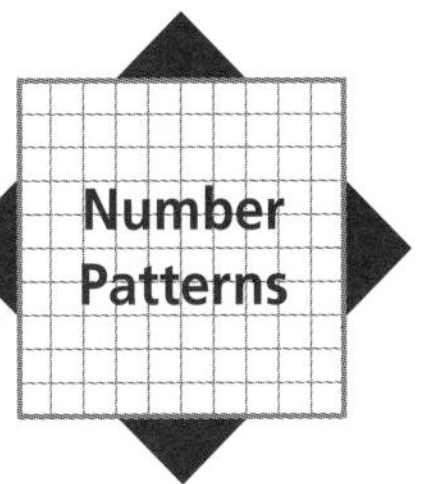

Worksheet 42

Even Numbers

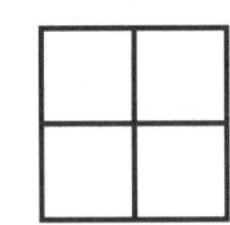 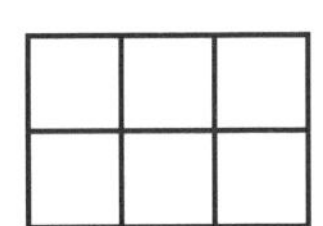 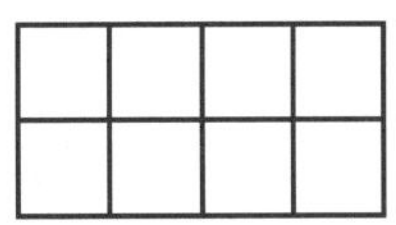

2 4 6 8 10

Odd Numbers

 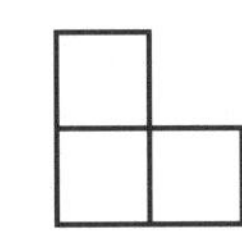 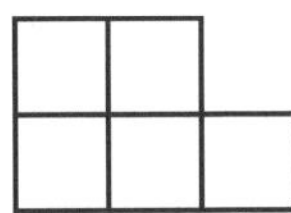 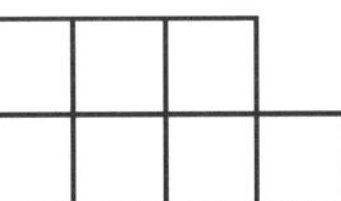 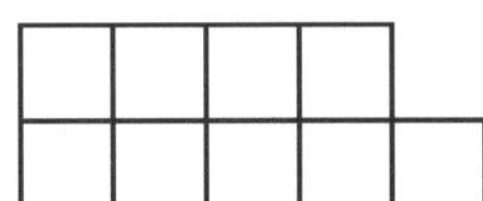 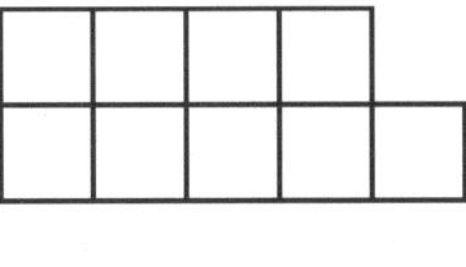

1 3 5 7 9

Each set of numbers below is even or odd.
Circle even or odd for each set.

1. 12, 14, 16, 18

even odd

2. 5, 7, 9, 11

even odd

3. 4, 6, 8, 10

even odd

4. 19, 21, 23, 25

even odd

5. 10, 20, 30, 40

even odd

6. 35, 37, 39, 41

even odd

Name ______________________________

Even and Odd

Worksheet 43

Number Patterns

Use the hundred board to answer these questions.

1. How many even numbers are on the board? _____

2. Circle the even numbers in this list.

   13 27 62 44 55

3. Write 4 different 2-digit even numbers.

   _____ _____ _____ _____

4. How many odd numbers are on the board? _____

5. What is the highest odd number? _____

6. Circle the odd numbers in this list.

   51 24 87 92 33

7. Write 4 different 2-digit odd numbers.

   _____ _____ _____ _____

8. Can you divide an odd number into 2 equal parts? _____

9. Will the numbers that come before and after every odd number be even? _____

Teacher's Notes

Activity 45

Addition to Ten

Materials
- Blank Hundred Board
- Twenty counters for each pupil, ten each of two colors

Students fill in the first row of the hundred board using both counter colors. Tally the number of each color to create an addition problem. For example, if 6 green and 4 blue counters fill the row, the addition problem becomes 6 + 4 = 10. Write the various problem combinations on the chalkboard if you wish.

0 + 10	1 + 9	2 + 8	3 + 7	4 + 6	5 + 5
10 + 0	9 + 1	8 + 2	7 + 3	6 + 4	

Activity 46

Moving Counters to Add

Materials
- Numbered Hundred Board
- Transparent counters
- Worksheet 46: Find the Sums

Demonstrate using the hundred board as an aid when solving addition problems. To find 7 + 3, children put one counter on the number 7. Next, they count 3 spaces and place a counter on the number 10. Show 7 + 5 as an example because children must count onto the second row to find the answer.

Worksheet 46 provides more addition practice on the hundred board.

Activity 47

Let's Add

Materials
- Two sets of ten index cards with numbers 1-10
- Colored counters
- Paper bag

Two players at a time will enjoy this game. Make two sets of ten index cards, numbered 1-10, then mix the cards together. Place the counters in a bag. Players take turns selecting two number cards and the corresponding number of counters. Students create number sentences with the counters, stating the problem and answering aloud. When players answer correctly, they keep the counters for that sum. Play until all counters have been divided. The player with the most counters wins.

Teacher's Notes

Activity 48

Add the Digits

Materials ● Numbered Hundred Board

Add each number of two-digit numbers together. If their sum is even, shade the number square on the hundred board. For example, with 11, 1 + 1 = 2. Two is an even number, so it should be shaded. However, with 12, 1 + 2 = 3, and 3 is not an even number, and should not be shaded.

Activity 49

Addition to Twenty

Materials ● Blank Hundred Board
● Twenty counters, ten each of two colors

Children cover the first row on their board using counters of one color. Use ten counters of a second color to add one at a time to the second row. As counters are added, form the following number sentences aloud:

10 + 1 = 11 10 + 4 = 14 10 + 7 = 17 10 + 9 = 19
10 + 2 = 12 10 + 5 = 15 10 + 8 = 18 10 + 10 = 20
10 + 3 = 13 10 + 6 = 16

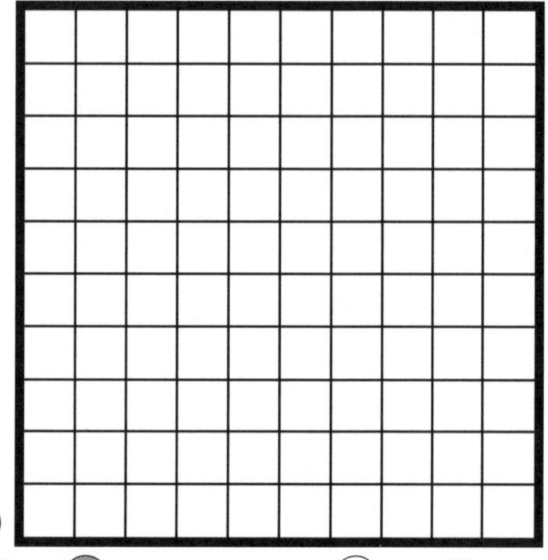

Activity 50

Patterns with Addition

Materials ● Numbered Hundred Board
● Crayons or colored pencils (nine different colors)

Create patterns with addition by following these instructions:

Color squares 1 to 9 with nine different colors. Use the color of square 1 to color in all the squares whose digits add up to 1 (10: 1 + 0 = 1). Use the color of square 2 to color in all squares whose digits add up to 2 (11, 20, etc.). When the sum of two digits is a two-digit number, as in 39, where 3 + 9 = 12, add again to get a one-digit number (1 + 2 = 3).

Continue in this manner for each of the colors and numbers up to 9.
Do students notice a pattern?

Teacher's Notes

Activity 51

Subtraction on the Hundred Board

Materials
- Blank Hundred Board
- Opaque or transparent counters

Students place ten counters in the first row on the hundred board. Next, take away one counter at a time, creating the following number sentences:

10 – 1 = 9	8 – 1 = 7	6 – 1 = 5	4 – 1 = 3	2 – 1 = 1
9 – 1 = 8	7 – 1 = 6	5 – 1 = 4	3 – 1 = 2	1 – 1 = 0

Next, replace all ten counters in the first row. Remove random numbers and create subtraction sentences 10 – 5 = 5.

Use more counters for difficult subtraction number sentences.

Activity 52-53

Let's Get 100

Materials
- Numbered Hundred Board
- Die numbered 1-6
- Two counters, different colors
- Worksheet 52: Two-Digit Addition
- Worksheet 53: Two-Digit Subtraction

1	2	3	4	5	6	7	8	9	10
11	12	13	14	15	16	17	18	19	20
21	22	23	24	25	26	27	28	29	30
31	32	33	34	35	36	37	38	39	40
41	42	43	44	45	46	47	48	49	50
51	52	53	54	55	56	57	58	59	60
61	62	63	64	65	66	67	68	69	70
71	72	73	74	75	76	77	78	79	80
81	82	83	84	85	86	87	88	89	90
91	92	93	94	95	96	97	98	99	100

Two players at a time will enjoy this adding game. Players start on square number 1 and take turns rolling the die. The number rolled indicates the number of spaces to move forward. Before moving, the player must state the addition sentence that represents the move. For example, a player who is on 1 and rolls a 4 says 1 + 4 = 5 before moving to 5. The first player to reach, or pass, 100 is the winner.

Repeat the game using subtraction. Start at 100 and work backward. Use Worksheets 52 and 53 for more addition and subtraction practice.

Activity 54

Addition of Two-Digit Numbers

Materials
- Numbered Hundred Board
- Counters

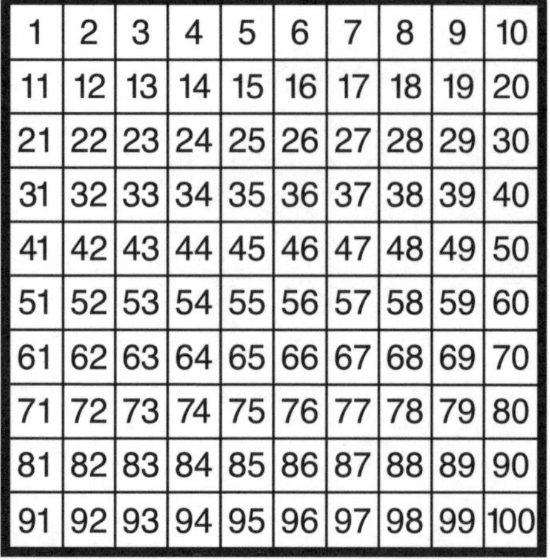

1	2	3	4	5	6	7	8	9	10
11	12	13	14	15	16	17	18	19	20
21	22	23	24	25	26	27	28	29	30
31	32	33	34	35	36	37	38	39	40
41	42	43	44	45	46	47	48	49	50
51	52	53	54	55	56	57	58	59	60
61	62	63	64	65	66	67	68	69	70
71	72	73	74	75	76	77	78	79	80
81	82	83	84	85	86	87	88	89	90
91	92	93	94	95	96	97	98	99	100

Introduce adding one- and two-digit numbers. Encourage students to use counters to keep their place on the hundred board. Demonstrate the following examples:

6 + 4 = 10
Start on 6.
Move forward 4.
Land on 10.

14 + 5 = 19
Start on 14.
Move forward 5.
Land on 19.

23 + 22 = 45
Start on 23.
Move down 2 rows.
Move forward 2 spaces.
Land on 45.

35 + 42 = 77
Start on 35.
Move down 4 rows.
Move forward 2 spaces.
Land on 77.

Activity 55

Subtraction of Two-Digit Numbers

Materials
- Numbered Hundred Board
- Counters

Introduce subtracting one- and two-digit numbers. Encourage children to use counters to keep their place on the hundred board. Demonstrate the following examples:

9 – 5 = 4
Start on 9.
Go back 5.
Land on 4.

18 – 3 = 15
Start on 18.
Go back 3.
Land on 15.

36 – 22 = 14
Start on 36.
Move up 2 rows.
Move back 2 spaces.
Land on 14.

54 – 28 = 26
Start on 54.
Move up 2 rows.
Move back 8 spaces.
Land on 26.

Name ____________________

Two-Digit Addition

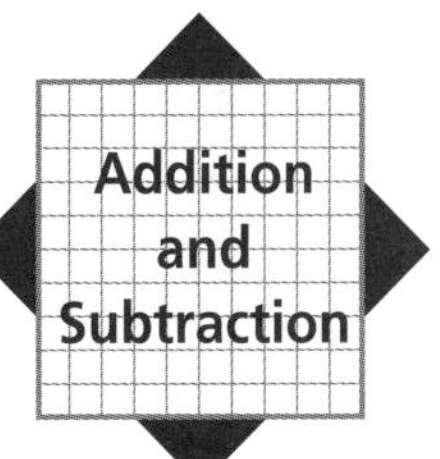

Worksheet 52

Use the hundred board to find the answers.

Start At	Move Forward	Total
5	5	
8	3	
12	4	
18	5	
20	6	
32	7	
53	8	
77	6	

Name ______________________

Two-Digit Subtraction

Addition and Subtraction

Worksheet 53

Use the hundred board to find the answers.

Start At	Move Back	Difference
15	5	
19	4	
26	3	
34	6	
48	7	
62	4	
79	5	

Name ______________________________

Fun with Arrows

Addition and Subtraction

Worksheet 58

 means to add

 means to subtract

Use the hundred board. Place a counter on the number shown. Move 1 space for each arrow. Write the number you land on.

1. 5 → → → → → ____

2. 9 → → → → → → ____

3. 18 → → → ____

4. 33 → → → → → ____

5. 35 ← ← ← ← ← ____

6. 12 ← ← ← ← ____

7. 56 ← ← ← ____

8. 84 ← ← ← ← ____

Teacher's Notes

Activity 62

Multiplication Charts for Twos

Name ________

Charts for Twos
Worksheet 62

Put counters on all the even numbers.
Then use the chart to help you multiply by 2.

1	2	3	4	5	6	7	8	9	10
11	12	13	14	15	16	17	18	19	20

1. 1 × 2 = ___
2. 2 × 2 = ___
3. 3 × 2 = ___
4. 4 × 2 = ___
5. 5 × 2 = ___
6. 6 × 2 = ___
7. 7 × 2 = ___
8. 8 × 2 = ___
9. 9 × 2 = ___
10. 10 × 2 = ___

58

Materials
- Numbered Hundred Board
- Worksheet 62: Charts for Twos

Ask pupils to lightly shade in every other number, starting with two. Then read the shaded boxes of the hundred board to complete the following number sentences:

1 × 2 = ______ 4 × 2 = ______ 7 × 2 = ______
2 × 2 = ______ 5 × 2 = ______ 8 × 2 = ______
3 × 2 = ______ 6 × 2 = ______ 9 × 2 = ______

Activity 63

Multiplication Charts for Threes

Name ________

Charts for Threes
Worksheet 63

Put counters on the 'threes.'
Then use the chart to help you multiply by 3.

1	2	3	4	5	6	7	8	9	10
11	12	13	14	15	16	17	18	19	20
21	22	23	24	25	26	27	28	29	30

1. 1 × 3 = ___
2. 2 × 3 = ___
3. 3 × 3 = ___
4. 4 × 3 = ___
5. 5 × 3 = ___
6. 6 × 3 = ___
7. 7 × 3 = ___
8. 8 × 3 = ___
9. 9 × 3 = ___
10. 10 × 3 = ___

59

Materials
- Numbered Hundred Board
- Worksheet 63: Charts for Threes

Ask pupils to lightly shade in every third number, starting with three. Then read the shaded boxes of the hundred board to complete the following number sentences:

1 × 3 = ______ 4 × 3 = ______ 7 × 3 = ______
2 × 3 = ______ 5 × 3 = ______ 8 × 3 = ______
3 × 3 = ______ 6 × 3 = ______ 9 × 3 = ______

Activity 64

Multiplication Charts for Fives

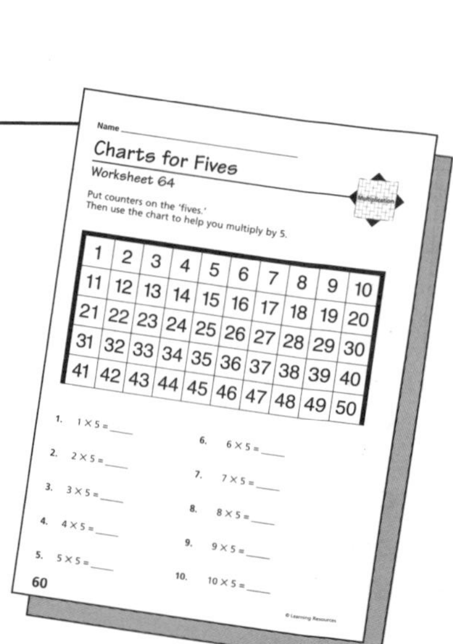

Name ________

Charts for Fives
Worksheet 64

Put counters on the 'fives.'
Then use the chart to help you multiply by 5.

1	2	3	4	5	6	7	8	9	10
11	12	13	14	15	16	17	18	19	20
21	22	23	24	25	26	27	28	29	30
31	32	33	34	35	36	37	38	39	40
41	42	43	44	45	46	47	48	49	50

1. 1 × 5 = ___
2. 2 × 5 = ___
3. 3 × 5 = ___
4. 4 × 5 = ___
5. 5 × 5 = ___
6. 6 × 5 = ___
7. 7 × 5 = ___
8. 8 × 5 = ___
9. 9 × 5 = ___
10. 10 × 5 = ___

60

Materials
- Numbered Hundred Board
- Worksheet 64: Charts for Fives

Ask pupils to lightly shade in every fifth number, starting with five. Then read the shaded boxes of the hundred board to complete the following number sentences:

1 × 5 = ______ 4 × 5 = ______ 7 × 5 = ______
2 × 5 = ______ 5 × 5 = ______ 8 × 5 = ______
3 × 5 = ______ 6 × 5 = ______ 9 × 5 = ______

Name ______________________________

Fun with Arrows

Worksheet 58

Addition and Subtraction

 means to add

 means to subtract

Use the hundred board. Place a counter on the number shown. Move 1 space for each arrow. Write the number you land on.

1. 5 ____

2. 9 ____

3. 18 ____

4. 33 ____

5. 35 ____

6. 12  ____

7. 56  ____

8. 84 ____

Name ______________________________

More Arrows

Addition and Subtraction

Worksheet 59

Means to add 1 ten.
Move down 1 space.

Means to subtract 1 ten.
Move up 1 space.

Move 1 space for each arrow.
Write the number you land on.

1. 15  ____

2. 30  ____

3. 23  ____

4. 49 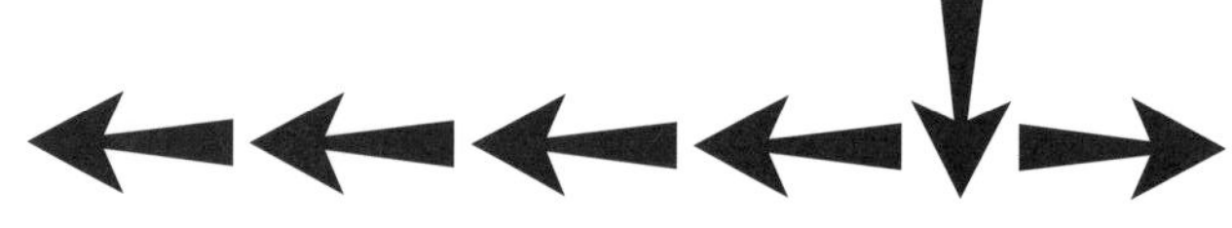 ____

5. 51  ____

6. 65  ____

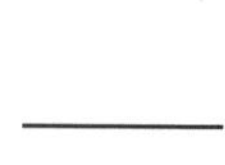

7. 77  ____

Teacher's Notes

Activity 60

Two by Two

Materials
- Blank Hundred Board
- Opaque or transparent counters

Instruct pupils to place two counters together on the hundred board. Then place another pair of counters in a different column. Explain that they are showing two groups of 2, which equals 4, and can be stated as 2 x 2 = 4.

Ask pupils to add two more counters to the hundred boards, showing three groups of 2 (2 x 3 = 6). Explain that multiplying by 2 is doubling a number, or adding a number to itself. Explain that multiplication is a shorter way to add numbers.

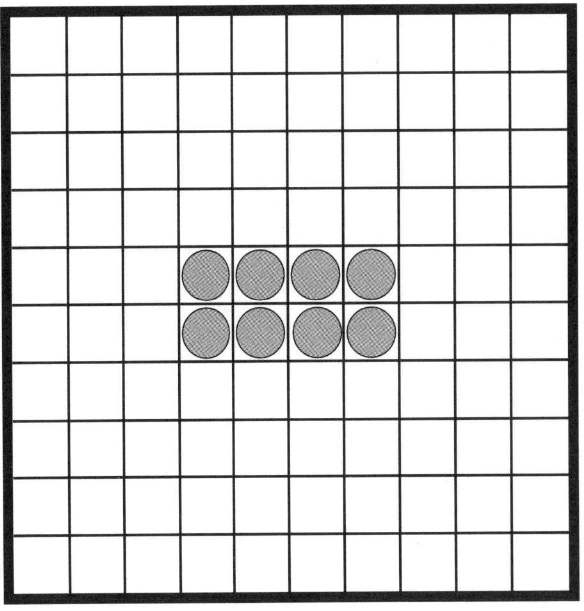

$2 \times 4 = 8$

$2 + 2 + 2 + 2 = 8$

Activity 61

Multiplication of Threes and Fives

Materials
- Blank Hundred Board
- Opaque or transparent counters

Instruct pupils to place three counters together (in different columns) on the hundred board. Do this three times. Ask pupils to give an addition number sentence for the counters shown (3 + 3 + 3 = 9). Ask how this could be written using multiplication (3 X 3 = 9).

Continue this exercise using examples of threes and fives, asking each time for the addition and multiplication equation.

Teacher's Notes

Activity 62

Multiplication Charts for Twos

Materials ● Numbered Hundred Board
● Worksheet 62: Charts for Twos

Ask pupils to lightly shade in every other number, starting with two. Then read the shaded boxes of the hundred board to complete the following number sentences:

1 × 2 = ______	4 × 2 = ______	7 × 2 = ______
2 × 2 = ______	5 × 2 = ______	8 × 2 = ______
3 × 2 = ______	6 × 2 = ______	9 × 2 = ______

Activity 63

Multiplication Charts for Threes

Materials ● Numbered Hundred Board
● Worksheet 63: Charts for Threes

Ask pupils to lightly shade in every third number, starting with three. Then read the shaded boxes of the hundred board to complete the following number sentences:

1 × 3 = ______	4 × 3 = ______	7 × 3 = ______
2 × 3 = ______	5 × 3 = ______	8 × 3 = ______
3 × 3 = ______	6 × 3 = ______	9 × 3 = ______

Activity 64

Multiplication Charts for Fives

Materials ● Numbered Hundred Board
● Worksheet 64: Charts for Fives

Ask pupils to lightly shade in every fifth number, starting with five. Then read the shaded boxes of the hundred board to complete the following number sentences:

1 × 5 = ______	4 × 5 = ______	7 × 5 = ______
2 × 5 = ______	5 × 5 = ______	8 × 5 = ______
3 × 5 = ______	6 × 5 = ______	9 × 5 = ______

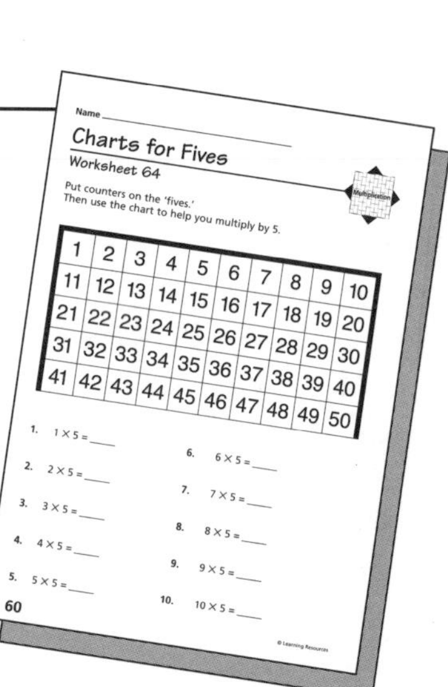

Name ______

Charts for Fives

Worksheet 64

Put counters on the 'fives.'
Then use the chart to help you multiply by 5.

1	2	3	4	5	6	7	8	9	10
11	12	13	14	15	16	17	18	19	20
21	22	23	24	25	26	27	28	29	30
31	32	33	34	35	36	37	38	39	40
41	42	43	44	45	46	47	48	49	50

1. 1 × 5 = ___
2. 2 × 5 = ___
3. 3 × 5 = ___
4. 4 × 5 = ___
5. 5 × 5 = ___
6. 6 × 5 = ___
7. 7 × 5 = ___
8. 8 × 5 = ___
9. 9 × 5 = ___
10. 10 × 5 = ___

60

Activity 65

Finding Multiples of Six

Materials
- Numbered Hundred Board
- Opaque counters

Instruct students to count to six, pointing at each number as they count. Cover square six with a counter. From square six, count six more squares, covering the last square (12). Continue counting and covering multiples of six in this manner.

Next, pupils can write the numbers they have covered. These numbers are the multiples of 6.

1	2	3	4	5		7	8	9	10
11		13	14	15	16	17		19	20
21	22	23		25	26	27	28	29	
31	32	33	34	35		37	38	39	40
41		43	44	45	46	47		49	50
51	52	53		55	56	57	58	59	
61	62	63	64	65		67	68	69	70
71		73	74	75	76	77		79	80
81	82	83		85	86	87	88	89	
91	92	93	94	95		97	98	99	100

Activity 66

Take Your Pick

Materials
- Numbered Hundred Board
- Number tiles 1-10
- Paper bag
- Fifteen counters per player, different colors for each player

This is a game for 2-4 players. Players take turns drawing three number tiles from the bag to form multiplication sentences. For example, if 2, 3, and 6 are drawn, the number sentence 2×3 or 2×6 or 3×6 could be used. The player decides on one number sentence and covers the product on the hundred board with a counter. The tiles are returned to the bag for the next player.

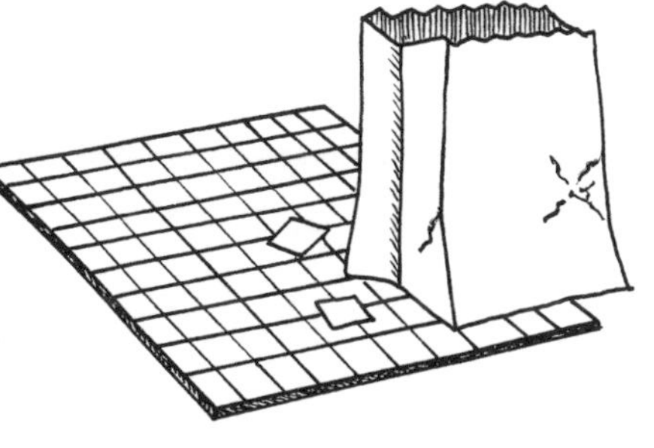

Players cannot share a number space. If all possible product spaces are filled, the player misses that turn and replaces the tiles. The first player to put all 15 counters correctly on the board wins.

Name ______________________________

Charts for Twos

Multiplication

Worksheet 62

Put counters on all the even numbers.
Use the chart to help you multiply by 2.

1	2	3	4	5	6	7	8	9	10
11	12	13	14	15	16	17	18	19	20

1. 1 × 2 = _____

2. 2 × 2 = _____

3. 3 × 2 = _____

4. 4 × 2 = _____

5. 5 × 2 = _____

6. 6 × 2 = _____

7. 7 × 2 = _____

8. 8 × 2 = _____

9. 9 × 2 = _____

10. 10 × 2 = _____

Name ________________________________

Charts for Threes

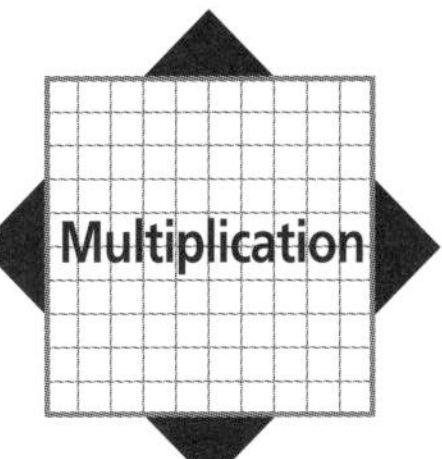

Put counters on the "threes." Use the chart to help you multiply by 3.

1	2	3	4	5	6	7	8	9	10
11	12	13	14	15	16	17	18	19	20
21	22	23	24	25	26	27	28	29	30

1. 1 × 3 = _____

2. 2 × 3 = _____

3. 3 × 3 = _____

4. 4 × 3 = _____

5. 5 × 3 = _____

6. 6 × 3 = _____

7. 7 × 3 = _____

8. 8 × 3 = _____

9. 9 × 3 = _____

10. 10 × 3 = _____

Name ______________________________

Charts for Fives

Multiplication

Worksheet 64

Put counters on the "fives."
Use the chart to help you multiply by 5.

1	2	3	4	5	6	7	8	9	10
11	12	13	14	15	16	17	18	19	20
21	22	23	24	25	26	27	28	29	30
31	32	33	34	35	36	37	38	39	40
41	42	43	44	45	46	47	48	49	50

1. 1 × 5 = _____

2. 2 × 5 = _____

3. 3 × 5 = _____

4. 4 × 5 = _____

5. 5 × 5 = _____

6. 6 × 5 = _____

7. 7 × 5 = _____

8. 8 × 5 = _____

9. 9 × 5 = _____

10. 10 × 5 = _____